Introduction to Wireless Sensor Networks

OrangeBooks Publication

1st Floor, Rajhans Arcade, Mall Road, Kohka, Bhilai, Chhattisgarh 490020

Website: **www.orangebooks.in**

© Copyright, 2024, Author

All rights reserved. No part of this book may be reproduced, stored in a retrieval system, or transmitted, in any form by any means, electronic, mechanical, magnetic, optical, chemical, manual, photocopying, recording or otherwise, without the prior written consent of its writer.

First Edition, 2024

INTRODUCTION TO
WIRELESS SENSOR NETWORKS

DR. SUHAS S. KHOT
PROF. RAHIMRAJA G. SHAIKH

OrangeBooks Publication
www.orangebooks.in

INDEX

UNIT – 1

Introduction To Wireless Sensor Networks 1

   1.1 WSN Overview 1

   1.2 Wireless Sensor Node 1

   1.3 Anatomy of a Sensor Node 3

   1.4 WSN Device Architecture and Operating System . 7

   1.5 Performance Metrics of WSNs 11

   1.6 Types of WSN 15

UNIT – 2

Wireless Communication and Link Management ... 26

   2.1 Properties of Wireless Communications 26

   2.2 Medium Access Protocols 30

   2.3 Wireless Links Introduction 36

   2.4 Metrics of Wireless Links 37

   2.5 Error Control 38

   2.6 Naming and Addressing 41

   2.7 Topology Control 44

UNIT - 3

Wireless Standards and Protocols............................ 48

   3.1 IEEE 802.15.4 Low Rate WPAN 48

   3.2 ZigBee... 50

   3.2.1 Wireless HART... 54

   3.2.2 ISA100.11a 55

   3.3. 6LoWPAN 57

   3.3.1 IEEE 802.15.3 60

   3.4. Wibree, BLE 62

   3.4.1 Z-Wave 63

   3.4.2 ANT 66

   3.4.3 INSTEON 68

   3.4.4 Wavenis 70

   3.5 Protocol Stack of WSNs 71

   3.6 Cross-Layer Protocols for WSNs........................ 80

UNIT - 4

Localization and Routing... 81

   4.1. Localization Challenges...................................... 81

   4.1.1 Types of Location Information 82

   4.1.2 Precision against Accuracy 83

   4.1.3 Costs... 83

   4.2 Pre-Deployment Schemes 84

   4.3 Proximity Schemes 84

4.4 Ranging Schemes 85

4.4.1 Range-Based Localization 88

4.4.2 Range-Free Localization 88

4.4.2.1 Hop-Based Localization 88

4.5. Routing Basics 90

4.5.1 Routing Metrics 91

4.5.1.1 Location and Geographic Vicinity 91

4.5.1.2. Hops 92

4.5.1.3 Number of Retransmissions 93

4.6 Routing Protocols 94

4.6.1 Full-Network Broadcasts 95

4.6.2 Location-Based Routing 95

4.6.3 Directed Diffusion 95

UNIT - 5

Data Aggregation and Security 96

5.1 Clustering Techniques 96

5.2 In-Network Processing and Data Aggregation 98

5.2.1 Compression 98

5.3 Security issues in Wireless Sensor Networks 100

5.4 Denial-of-service attacks 101

5.5 Sensor Network Security properties 103

UNIT - 6

Designing and Deploying WSN Applications 105

6.1 Early WSN Deployments 105

6.1.1 Murphy Loves Potatoes 105

6.1.2 Great Duck Island 106

6.2 General Problems 108

6.2.1 Node Problems 109

6.2.2 Link/Path Problems 111

6.2.3 Global Problems 112

6.3 General Testing and Validation 115

6.4 Requirements Analysis 117

6.5 The Top-Down Design Process 122

6.6 Bottom-Up Implementation Process 127

•••

UNIT - 1

Introduction To Wireless Sensor Networks

1.1 WSN Overview

A sensor node is a node in a sensor network that is capable of performing some processing, gathering sensed information and communicating with other connected nodes in the network. The main components of a sensor node are a microcontroller, transceiver, external memory, power source and one or more sensors.

1.2 Wireless Sensor Node

A wireless sensor network (WSN) is an association of spatially dispersed independent sensor nodes. The sensors in each node are capable of sensing and monitoring physical or environmental parameters, such as temperature, sound, vibration, pressure, motion or pollutants in different areas. The data obtained by each sensor will then cooperatively pass through the network to a main control station. The main control station will take decisions based on the received data. Figure 1.1 shows one of the wireless sensor network topology.

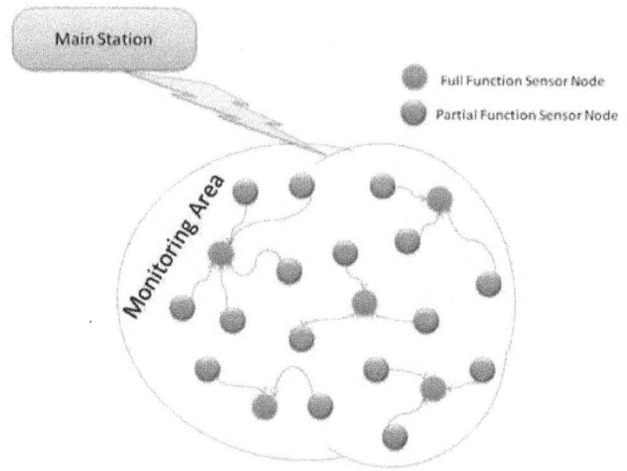

Figure 1.1: Structural Design of Wireless Sensor Network

WSNs are widely used for several military, industrial, civilian, environmental and medical applications. In all these applications, the sensor nodes are deployed over a region where some phenomenon or some parameters are to be monitored. Wireless sensor nodes make the task more mobile for sensing and collecting the data from different areas. Wireless sensor networks are deployed in many countries to monitor air pollution, water pollution, landslides, floods and infrastructure of a building. Size and cost constraints on sensor nodes result in corresponding constraints on resources such as energy, memory, computational speed and communications bandwidth.

The topology of the WSNs can vary from a simple star network to an advanced multi-hop wireless mesh network or from a centralized network to a hierarchical network.

To make the wireless sensor network energy optimized, and to establish effective communication between the main station and the sensor nodes, the sensor nodes are classified into full function nodes and partial function nodes. Full function sensor nodes have additional features which will collect data from the nearby partial function nodes, aggregate it and send the data to the main station. Such architecture is shown in the figure 1.1.

1.3 Anatomy of a Sensor Node

Sensing is what distinguishes the living from stones and rocks. Alive creatures have levels and ways of sensing, without sensing there is no communication with the outside word, there is no life. Lecturing on zoology or botany is not an objective, but a quick reminder on senses of the living is recalled (Birds and Blooms 2013). Many animals see the world completely differently to humans. Being able to see helps animals locate food, move around, find mates and avoid predators, whether they live at the bottom of the ocean or soar high in the sky. Eyesight is important foremost animals and nearly all animals can see, 95 % of all species have eyes. Some animals live in complete darkness in caves or underground, where they cannot see anything, their eyes often no longer work, but they have developed an extra-sensitive sense of touch to feel their way around.

However, only two animal groups have evolved the ability to hear, vertebrates like mammals, birds and reptiles, and arthropods, such as insects, spiders and crabs. No other animals can hear. Some animals have a remarkable sense of hearing, finely tuned to where and how they live, many animals hear sounds that humans cannot. Human senses of smell and taste are feeble compared to those of many other animals, a keen sense of smell allows animals to find food and mates, as well as to stay out of danger, it can stop an animal wandering into a rival's territory or help it find its way.

Animals communicate using visual signals, sounds, touch, smells and tasks. Vision, touch and taste work well over short distances, but sounds travel much further and scent marks can last long after the animal has moved on. Sometimes the aim is to deceive, blending into the background, pretending to be a twig or playing dead; animals give out all sorts of false information to avoid danger or help catch their next meal. Their tricks and deceptions vary from camouflage and mimicry to distracting, startling, scaring and confusing others (National Museum Scotland 2013).

An insect's acute sense of smell enables it to find mates, locate food, avoid predators, and even gather in groups. Insects have sense organs for taste, touch, smell, hearing, and sight. Some insects have sense organs for temperature and humidity as well as stresses and movements of their body parts. Some insects rely on chemical cues to find their way to and from a nest, or to space themselves appropriately in a habitat with limited resources. Insects, you may have noticed, do not have noses. So how are they able to sense the faintest of scents in the wind? Antennae sometimes are called "feelers".

However, antennae as primarily "smellers "are the insect's noses because they are covered with many organs of smell. These organs help the insect to find food, a mate, and places to lay eggs. Insect seven can decide which directions to fly by using their sense of smell (O. Orkin Insect Zoo 2013). How do fish sense movement? Fish have the five senses that people have, but have a sixth sense that is more than a sense of touch. Fish have a row of special cells inside a special canal along the surface of the fish's skin. This is called the "lateral line" which allows them to detect water vibrations. This sixth sense allows fish to detect movement around them and changes in water flow.

Detecting movement helps fish find prey or escape from predators. Detecting changes in water flow help fish chose where to swim (Texas Parks and Wildlife 2013). What about birds? They depend less on the senses of smell and taste than people do. The odors of food, prey, enemies or mates quickly disperse in the wind. Birds possess olfactory glands, but they are not well developed in most species, including the songbirds in our backyards. The same is true for taste, which is related to smell. While humans have 9,000 taste buds, songbirds have fewer than 50.

That means the birds we feed around must locate their food by sight or touch, two senses that are highly developed in birds (Birds and Blooms 2013). Plants, unlike animals, do not have ears, eyes, or tongues to help them feel and acquire information from their environment. But without being helpless, they do sense their environment in other ways and respond accordingly. Plants can detect various wavelengths and use colors to tell them what the environment is like. When a plant grows in the shadow of another, it will send a shoot straight up towards the light source, it has also been shown that plants know when it is day and when it is night. Leaf pores on plants open up to allow photosynthesis during the daytime and close at night to reduce water loss. Plants also respond to ultraviolet light by producing a substance that is essentially a sunscreen so that they do not get sunburned. They can sense weather changes and temperatures as well. Plants have specific regulators, plant hormones, minerals and ions that are involved in cell signaling and are important in environmental sensing.

In fact, without these, the plants will not grow properly (UCSB Science Line 2013). Reminding of human senses is easy, the use of eye contacts, the eye attraction to what is beautiful, the love of perfumes, the appreciation of beautiful music, there living touch of softness, the tantalizing taste of sweeties. It is all senses. Human interaction with the environment is an eternal task that grows and expands with expansion of ambitions, with technology. This book is interested in presenting wireless sensor networks (WSNs) in comprehensive details that are far beyond what birds, insects, mammal scan. As an opening start, the goal of this chapter is to present a thorough survey of WSNs.

Mobile Ad Hoc Networks (MANETs):

At first it is needed to strengthen up basics, a Mobile Ad hoc NET work (MANET)is one that comes together as needed, not necessarily with the support of an existing Internet infrastructure or any fixed station, it is an autonomous system of mobile hosts serving as routers and connected by wireless links (Cordeiro and Agrawal2002). This contrasts the single hop cellular network that supports the need for wireless communication by installing base stations as access points, such that the communication between wireless nodes rely on the wired backbone and the fixed base stations. In a MANET there is no infrastructure and the network topology changes unpredictably since nodes are free to move. As for the mode of operation, ad hoc networks are peer-to-peer multi-hop mobile wireless networks where information packets are transmitted in a store and forward manner from source to destination via intermediate nodes as shown in Figure. 1.2. Topology changes as.

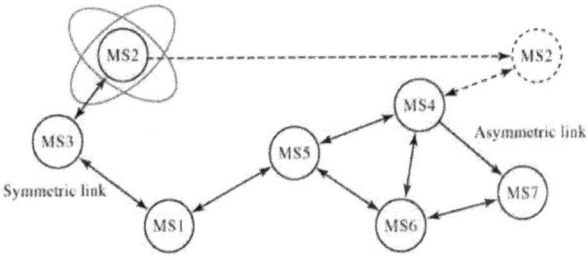

Figure 1.2: Mobile Ad hoc Network

1.4 WSN Device Architecture and Operating System

WSNs are widely used for several military, industrial, civilian, environmental and medical applications. In all these applications, the sensor nodes are deployed over a region where some phenomenon or some parameters are to be monitored. Wireless sensor nodes make the task more mobile for sensing and collecting the data from different areas. Wireless sensor networks are deployed in many countries to monitor air pollution, water pollution, landslides, floods and infrastructure of a building. Size and cost constraints on sensor nodes result in corresponding constraints on resources such as energy, memory, computational speed and communications bandwidth. The topology of the WSNs can vary from a simple star network to an advanced multi-hop wireless mesh network or from a centralized network to a hierarchical network. To make the wireless sensor network energy optimized, and to establish effective communication between the main station and the sensor nodes, the sensor nodes are classified into full function nodes and partial function nodes. Full function sensor nodes have additional features which will collect data from the nearby partial function nodes, aggregate it and send the data to the main station. Such architecture is shown in the figure 1.3.

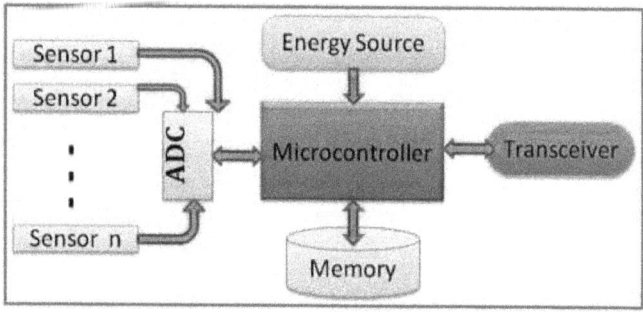

Figure 1.3: Architecture of wireless sensor node

The sensor nodes that create the WSN should be of low cost and small size. These two aspects are considered as the major challenges in the development of sensor nodes. The design of the operating systems for the sensor nodes can be of two types, namely, event-driven and multithreading.

- In event driven method, every action of the operating system is triggered by an event while in multithreading, each task is considered as threads and multiplex the execution time between different threads.
- Tiny Os is an example of the event driven operating system and MANTIS is an example of the multithreaded operating system for sensor nodes.

The types of wireless sensor nodes that are currently commercially available are Barkley motes and its replicas. Some of the commonly used motes are Iris motes, I Sense motes, Micas motes and WASP motes. Much research is going on in the development of energy optimized and cost effective small sized motes.

Architecture of WSNs:

The term architecture has been adopted to describe the activity of designing any kind of system, it is the complex or carefully designed structure of something; one of its common uses is in describing information technology, such as computer architecture and network architecture. The architecture of WSNs is built up of main entities as shown in Fig. 1.4:

- The Sensor nodes that form the sensor network. Their main objectives are making discrete, local measurement about phenomenon surrounding these sensors, forming a wireless network by communicating over a wireless medium, and collecting data and routing data back to the user via a sink (base station).

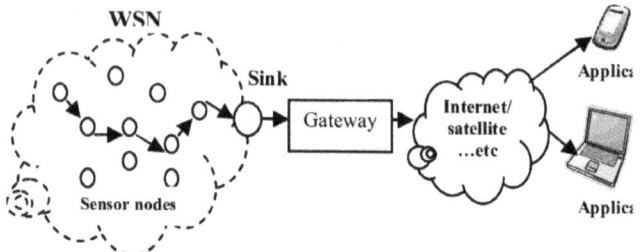

Figure 1.4: Architecture of WSNs

The sink (base station) communicates with the user via Internet or satellite communication. It is located near the sensor field or well-equipped nodes of the sensor network. Collected data from the sensor field is routed back to the sink by a multi-hop infra-structure less architecture through the sink.

- The phenomenon which is an entity of interest to the user, it is sensed and analyzed by the sensor nodes.

- The user who is interested in obtaining information about a specific phenomenon onto measure/monitor its behavior. Although many protocols and algorithms have been proposed for traditional wireless ad hoc networks, they are not well suited for the unique features and application requirements of sensor networks, as detailed in this section. For further illustration, the differences between WSNs and MANETs are outlined below (Akyildiz et al. 2002a):

- The number of sensor nodes in WSNs can be several orders of magnitude higher than the nodes in MANETs.

- Sensor nodes are densely deployed.

- Sensor nodes are prone to failures.

- The topology of a sensor network changes very frequently.

- Unlike a node in MANETs, a sensor node may not have a unique global IP address due to the numerous numbers of sensors and the resulting high overhead.

- Sensor nodes as deployed in high numbers, are extremely cheap and considerably tiny, unlike MANET nodes (e.g., PDAs, Laptops, etc.).

- The communication paradigm used in WSNs is broadcasting, whereas MANETs are based on point-to-point communications.

- The topology of a WSN changes very frequently.

- Limited energy and bandwidth conservation is the main concern in WSN protocols design, which is not really worrisome in MANETs.

1.5 Performance Metrics of WSNs

Metric is the standard of measurement; it varies with the measured environment. Time delay is a widely used metric, it is the time needed to obtain a response after applying certain input, its units are coarsely seconds, but specifically at which scale? In an electronic environment, time delay units are microseconds and less, in electro-mechanical environment they are milliseconds or more, in pure mechanical systems they are seconds and above. In athletic run sports, speed is the metric, its unit scale varies with distances, from the 100 m race till the marathon. Generally, speed varies with who is running and where, a professional human runner spends 2:15 h in a

42.195 km marathon, while a cheetah that is three times faster just needs 25 min to reach the end point; the metric is time, the same, but for humans it measured in hours, while it is in minutes for the cheetah (Fig. 1.5). One of the metrics for goods is weight, its unit is kilograms or pounds, but for coal it is a multiplicity of kilograms for home use and tons for industry. On the other hand, gold weight is calculated in grams or ounces for personal use and in kilograms for

gold traders Lifetime a metric related to living being existence, it is left for the reader to have some thoughts about it, at least for mind relief. Back to WSNs, for a WSN to perform satisfactorily, some metrics are also defined, measured, and interpreted far from confusion. Several metrics, close to WSNs characteristics as introduced in previous sections of this chapter, evaluate sensor network performance. Specifically:

- **Network lifetime**: It is measure of energy efficiency, as sensor nodes are battery operated, WSNs protocols must be energy efficient to maximize system lifetime. System lifetime can be measured by generic parameters such as the time until half of the nodes die, or by application directed metrics, such as when the network stops providing the application with the desired information about the environment, it is also calculated as the time until message loss rate exceeds a given threshold.

- **Energy consumption:** It is the sum of used energy by all WSN nodes. The consumed energy of a node is the sum of the energy used for communication, including transmitting, receiving, and idling. Assuming each transmission consumes an energy unit, the total energy consumption is equivalent to the total number of packets sent in the network.

Figure 1.5: Fastest runners with different metrics
  a) Usian Bolt hits 9:58 s for 100 m
  b) Cheetah fastest runner on earth

- **Latency:** It is the end-to-end delay that implies the average time between sending a packet from the source, and the time for successfully receiving the message at the destination. Measurement takes into account the queuing and the propagation delay of the packets. The observer is interested in getting information about the environment within a given delay. The precise units of latency are application dependent.

- **Accuracy**: It is the freedom from mistake or error, correctness, conformity to truth, exactness. Obtaining accurate information is the primary objective of the observer. There is a trade-off between accuracy, latency and energy efficiency. A WSN should be adaptive such that its performance achieves the desired accuracy and delay with minimal energy expenditure. For example, the WSN task, the application, can either request more frequent data dissemination from the same sensor nodes, or it can direct data dissemination from more sensor nodes with the same frequency

- **Fault-tolerance:** Sensors may fail due to surrounding physical conditions or when their energy runs out. It may be impractical to replace existing sensors; in response, the WSN must be fault- tolerant such that non-serious failures are hidden from the application in a way that does not hinder it. Fault-tolerance maybe achieved through data replication, as in the SPIN protocol (Xiao et al. 2006). However, data replication itself requires energy; there is a trade-off between data replication and energy efficiency, generally, data replication should be application-specific, higher priority data according to the application might be replicated for fault-tolerance.

- **Scalability:** As a prime factor, it is WSN adaptability to increased workload that is to include more sensor nodes than what was anticipated during network design. A scalable network is one that can be expanded in terms of the number of sensors, complexity of the network topology, data quality, e.g., sampling rate, sensor sensitivity, and amount of data while the cost of the expansion installation and operational cost, communication time, processing time, power, and reliability is no worse than a linear, or nearly linear, function of the number of sensors (Pakzad et al. 2008). WSN scalability needs to consider an integrated view of the hardware and software. For hardware, scalability involves sensitivity and range of MEMS sensors, communication bandwidth of the radio, and power usage. The software issues include reliability of command dissemination and data transfer, management of large volume of data, and scalable algorithms for analyzing the data. The combined hardware-software issues include high-frequency sampling, and the tradeoffs between on-board computations compared with wireless communication between nodes.

- **Network throughput:** It is a common metric for all networks. The end-to-end throughput measures the number of packets per second received at the destination.

1.6 Types of WSN

WSNs can be deployed on ground, underground, and underwater. Five functional types can be distinguished, pacifically, terrestrial, underground, underwater, multi-media, and mobile WSNs (Yick et al. 2005). What follows provides the details of each type.

1. Terrestrial WSNs:

Terrestrial WSNs deployed in a given area (Yick et al. 2008). There are two ways to deploy sensor nodes on WSNs:

- In Unstructured WSN, this contains a dense collection of sensor nodes. Sensor nodes may be deployed in an ad hoc manner into the field, once deployed the network is left unattended to perform monitoring and reporting functions. In an unstructured WSN, network maintenance such as managing connectivity and detecting failures is difficult since there are so many nodes.

- In Structured WSN, all or some of the sensor nodes are deployed in a pre-planned manner. The advantage of a structured network is that fewer node scan be deployed with lower network maintenance and management cost. Fewer nodes are beneficially deployed since they are placed at specific locations to provide coverage while ad hoc deployment can have uncovered regions. Sensor nodes are deployed on the sensor field within reach of the transmission range of each other and at densities that may be as high as 20 nodes/m3. Densely deploying hundreds or thousands of sensor nodes over a field requires maintenance of topology along three phases.

- Pre-deployment and deployment phase. Sensor nodes may either be thrown in the deployment field as a mass from an airplane or an artillery shell, or place done by one by a human or a robot.

- Post deployment phase. After deployment, topology changes due to change in sensor nodes position, reach ability (that may be effected by jamming, noise, moving obstacles, etc.), remaining energy, malfunctioning, and task details.

- Redeployment of additional nodes. Additional sensor nodes can be re deployed to replace malfunctioning nodes or to account for changes in task dynamics. In a terrestrial WSN, reliable communication in a dense environment is a must. Sensor nodes must be able to effectively communicate data back to the base station. While battery power is limited and may not be rechargeable, terrestrial sensor node show ever can be equipped with a secondary power source such as solar cells, it is important for sensor nodes to conserve energy. For a terrestrial WSN, energy can be conserved with multi-hop optimal routing, short transmission range, in-network data aggregation, laminating data redundancy, minimizing delays, and using low duty-cycle operations.

2. Underground WSNs:

Underground WSNs consist of a number of sensor nodes buried underground or in a cave or mine used to monitor underground conditions (Li and Liu 2007, 2009; Liet al. 2007). Additional sink nodes are located above ground to relay information from the sensor nodes to the base station. An underground WSN is more expensive than a terrestrial WSN in terms of equipment, deployment, and maintenance. Underground sensor nodes are expensive because appropriate equipment parts must be selected to ensure reliable communication through soil, rocks, water, and other mineral contents. The underground environment makes wireless communication a challenge due to signal losses and high levels of attenuation. Unlike terrestrial WSNs, the deployment of an underground WSN requires careful planning and energy and cost considerations. Energy is an important concern in underground WSNs. Like terrestrial WSN, Underground sensor nodes are equipped with a limited battery power and once deployed into the ground, it is difficult to recharge or replace a sensor node's battery. As usual, a key objective is to conserve energy in order to increase the network lifetime, which can be achieved by implementing efficient communication protocol.

3. Underwater Acoustic Sensor Networks (UASNs):

Underwater acoustic sensor networks (UASNs) technology provides new opportunities to explore the oceans, and consequently it improves understanding of the environmental issues, such as the climate change, the life of ocean animals and the variations in the population of coral reefs. Additionally, UASNs can enhance the underwater warfare capabilities of the naval forces since they can be used for surveillance, submarine detection, mine countermeasure missions and unmanned operations in the enemy fields. Furthermore, monitoring the oil rigs with UASNs can help taking preventive actions for the disasters such as the rig explosion that took place in the Gulf of Mexico in 2010. Last but not least, earthquake and Tsunami forewarning systems can also benefit from the UASN technology (Erol-Kantarci et al. 2011).

Ocean monitoring systems have been used for the past several decades, where traditional oceanographic data collection systems utilize individual and disconnected underwater equipment. Generally, this equipment collects data from their surroundings and sends these data to an on-shore station or a vessel by means of satellite communications or underwater cables. In UASNs, this equipment is replaced by relatively small and less expensive underwater sensor nodes that house various sensors on board, e.g., salinity, temperature, pressure, current speed sensors. The underwater sensor nodes are networked, unlike the traditional equipment, and they communicate underwater via acoustics. In underwater, radio signals attenuate rapidly, hence they can only travel to short distances while optical signals scatter and cannot travel far in adverse conditions, as well. On the other hand, acoustic signals attenuate less, and they are able to travel further distances than radio signals and optical signals.

Consequently, acoustic communication emerges as a convenient choice for underwater communications. However, it has several challenges (Heidemann et al. 2006): The bandwidth of the acoustic channel is low, hence the data rates are lower than they are in terrestrial WSNs. Data rates can be increased by using short range communications, which means more sensor nodes will be required to attain a certain level of connectivity and coverage. In this respect, large-scale UASN poses additional challenges for communication and networking Protocols.

- The acoustic channel has low link quality, which is mostly due to the multi-path propagation and the time-variability of the medium.

- Furthermore, the speed of sound is slow (approximately 1500 m/s) yielding large propagation delay.

- In mobile UASNs, the relative motion of the transmitter or the receiver may create the Doppler effect.

- UASNs are also energy limited similar to other WSNs. Due to the above challenges, UASNs rooms research studies in novel medium access, network, transport, localization, synchronization protocols and architectures (Jornet et al. 2008; Vuran and Akyildiz2008; Lee et al. 2010; Ahna et al. 2011). The design of network and management protocols is closely related with the network architecture, and various UASN architectures have been proposed in the literature. Moreover, localization has been widely addressed since it is a fundamental task used in tagging the collected data, tracking underwater nodes, detecting the location of an underwater target and coordinating the motion of a group of nodes, Furthermore, location information can be used to optimize the medium access and routing protocols (Chandrasekhar et al. 2006; Erol-Kantarci et al. 2011;Zhou et al. 2011).Underwater sensor nodes must be able to self-configure and adapt to harsh ocean environment, they are equipped with a limited battery, which cannot be replaced or recharged. The issue of energy conservation for underwater WSNs involves developing efficient underwater communication and networking techniques.

4. Multimedia WSNs:

Multimedia WSNs have been proposed to enable monitoring and tracking of events in the form of multimedia such as video, audio, and imaging (Akyildiz et al. 2007). Multimedia WSNs consist of a number of low-cost sensor nodes equipped with cameras and microphones. These sensor nodes interconnect with each other over a wireless connection for data retrieval, processing, correlation, and compression. They are deployed in a preplanned manner into the environment to guarantee coverage. Challenges in multimedia WSN include:

- High bandwidth demand.
- High energy consumption.
- Quality of service (QoS) provisioning.
- Data processing and compressing techniques.
- Cross-layer design.

Multimedia content such as a video stream requires high bandwidth in order for the content to be delivered quickly, consequently, high data rate leads to high energy consumption. Thus, transmission techniques that support high bandwidth and low energy consumption have to be developed. QoS provisioning is a challenging task in a multimedia WSN due to the variable delay and variable channel capacity. It is important that a certain level of

QoS must be achieved for reliable content delivery. In-network processing, filtering, and compression can significantly improve network performance in terms of filtering and extracting redundant information and merging contents. Similarly, cross-layer interaction among protocol layers can improve the processing and delivering of data.

5. Mobile WSNs:

Mobile WSNs consist of a collection of sensor nodes that can move on their own and interact with the physical environment (Francesco et al. 2011). There are several comparative issues between mobile and static sensor nodes: Like static nodes, mobile nodes have the ability to sense, compute, and communicate. Contrarily, mobile nodes have the ability to reposition and organize themselves in the network. A mobile WSN can start off with some initial deployment and nodes can then spread out to gather information. Information gathered by a mobile node can be communicated to another mobile node when they are within range of each other. Another key difference is data distribution. In a static WSN, data can be distributed using fixed routing or flooding while dynamic routing is used in a mobile WSN. Mobility in WSNs is useful for several reasons, as presented in what follows (Anastasia et al. 2009):

- **Connectivity**: As nodes are mobile, a dense WSN architecture is not a pressing requirement. Mobile elements can cope with isolated regions, such that the constraints on network connectivity and on nodes (re) deployment can be relaxed. Hence, a sparse WSN architecture becomes a feasible option.

- **Cost**: Since fewer nodes can be deployed, the network cost is reduced in a mobile WSN. Although adding mobility features to the nodes might be expensive, it may be possible to exploit mobile elements, which are already present in the sensing area (e.g., trains, buses, shuttles or cars), and attach sensors to them.

- **Reliability:** Since traditional (static) WSNs are dense and the communication paradigm is often multi-hop ad hoc, reliability is compromised by interference and

- Collisions moreover, message loss increases with the increase in number of hops. Mobile elements, instead, can visit nodes in the network and collect data directly through single-hop transmissions; this reduces not only contention and collisions, but also the message loss.

- **Energy efficiency:** The traffic pattern inherent to WSNs is converge cast, i.e., messages are generated from sensor nodes and are collected by the sink. As a consequence, nodes closer to the sink are more overloaded than others, and subject to premature energy depletion. This issue is known as the funneling effect, since the neighbors of the sink represent the bottleneck of traffic. Mobile elements can help reduce the funneling effect, as they can visit different regions in the network and spread the energy consumption more uniformly, even in the case of a dense WSN architecture. However, mobility in WSNs also introduces significant challenges, which do not arise in static WSNs, as illustrated below:

- **Contact detection:** Since communication is possible only when the nodes are in the transmission range of each other, it is necessary to detect the presence of a mobile node correctly and efficiently. This is especially true when the duration of contacts is short.

- **Mobility-aware power management:** In some cases, it is possible to exploit the knowledge on the mobility pattern to further optimize the detection of mobile elements. In fact, if visiting times are known or can be predicted with certain accuracy, sensor nodes can be awake only when they expect the mobile element to be in their transmission range.

- **Reliable data transfer:** As available contacts might be scarce and short, there is a need to maximize the number of messages correctly transferred to the sink. In addition, since nodes move during data transfer, message exchange must be mobility-aware.

- **Mobility control:** When the motion of mobile elements can be controlled, a policy for visiting nodes in the network has to be defined. That is, the path and the speed or sojourn time of mobile nodes have to be defined in order to improve (maximize) the network performance.

- **Challenges:** include deployment, localization, navigation and control, coverage, maintenance, and data processing. Mobile WSN applications include environment monitoring, target tracking, search and rescue, and real-time monitoring of hazardous material. Mobile sensor nodes can move to areas of events after deployment to provide the required coverage. In military surveillance and tracking, they can collaborate and make decisions based on the target. Mobile sensor nodes can achieve a higher degree of coverage and connectivity compared to static sensor nodes. In the presence of obstacles in the field, mobile sensor nodes can plan ahead and move appropriately to unobstructed regions to increase target exposure.

•●•

UNIT - 2

Wireless Communication and Link Management

Wireless Communication Overview: Wireless Communication (or just wireless) is the transfer of information between two or more points without the use of any physical medium for the transfer. The most common wireless technologies use radio waves.

2.1 Properties of Wireless Communications

While traveling through the environment the electromagnetic wave experiences multiple distortions. These are mainly due to the following processes, which are shown in Figure 2.1.

- **Attenuation.** This process spreads the energy of the wave to larger space. It is similar to a balloon, which is a dark red color before filling it with air, but then becomes almost transparent once filled. Thus, with growing distance from the sender, the wave becomes less and less powerful and harder to detect (Figure 2.1a).

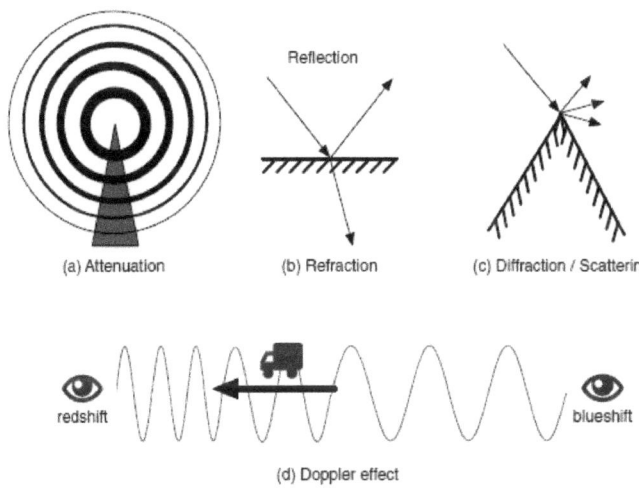

Figure 2.1: Different physical processes that lead to path loss in signal travelling.

- **Reflection/Refraction.** This process changes the direction of the wave when it meets a surface. Part of the wave gets reflected and travels a new trajectory; another part of the wave gets refracted into the material and changes its properties. Both processes create new, secondary waves, which also reach the receiver at some point in time, slightly after the primary wave. This is both a blessing and a curse-very weak signals can be better detected when the primary and the secondary signals overlap. At the same time, it is hard to understand what a secondary wave versus a primary wave is (Figure 2.1b).

- **Diffraction/Scattering.** Sharp edges and uneven surfaces in the environment can break the wave into several secondary waves with the same consequences as above (shown in Figure 2.1c).

- **Doppler effect.** In general, the frequency of the signal changes with its relative velocity to the receiver. The Doppler effect is well known for its impact on the police siren, which sounds different to the observer depending on whether the police car is approaching or moving away. The same happens with the radio waves when their frequencies get shifted in one or the other direction which results in a loss of center. For wireless communications, this means that you can hardly differentiate between various frequency codes (Figure 2.1d).

Definition of Path Loss: Path loss is the reduction in power density of an EM wave as it travels through space.

Path loss is central to wireless communications, as it allows you to predict the quality of the transmission and/or to design wireless links. In the remainder of this section, we will explore how path loss behaves in reality and impacts wireless communications.

Interference and Noise:

So far, this chapter has only considered wireless transmission problems resulting from a single sender and its environment. However, typically there are several simultaneous senders and other even non-manmade sources of disturbance. The latter is also referred to as noise.

Definition of Electromagnetic Noise: Electromagnetic Noise is the unwanted fluctuation in a signal from natural sources.

On the other hand, the corruption of a signal through other active senders is called interference.

Definition of Electromagnetic Interference:

Electromagnetic interference is the disturbance of EM Signal due to some external source.

Hidden Terminal Problem:

The hidden terminal problem is best explained with a diagram (Figure 2.5). As you can see, there are four nodes present. Node A starts sending packet X to node B Node C is situated outside of the transmission range of node A, so it does not know anything about the ongoing transmission of packet X. Transmission range is generally referred to as a semi- circular area around a sender where an ongoing transmission can be detected.

Since node C does not know anything about the ongoing transmission between A and B, it starts to transmit a packet to node D. This causes interference at node B, which corrupts packet X. However, the transmission between C and D is successful.

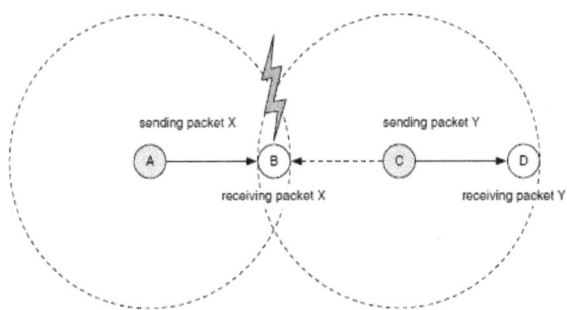

Figure 2.2: The hidden terminal problem in wireless communications.

Node A tries to send packet X to node B. At the same time, node C decides to send to node D, because it does not hear anything from node A (out of transmission range). The interference of both packets at node B results in packet loss. However, the packet at node D is received.

The hidden terminal problem is a major challenge to overcome in wireless communications. Later in this chapter in Section 3.3.3, you will see how to resolve it and to let neighboring nodes know about ongoing transmissions.

Exposed Terminal Problem:

The exposed terminal problem is the opposite of the hidden terminal problem. Figure 2.3 illustrates the case, when a node is prevented to send a packet because of another ongoing transmission. This time, node B is sending a packet to node A. At the same time, node C has a packet to send to node D. Looking at the topology of the network and transmission ranges, you can see that both transmissions are simultaneously possible. However, node C cannot send a packet because it hears the ongoing transmission of node B and does not know where node A is. Thus, it needs to wait, even if the transmission is possible.

2.2 Medium Access Protocols

The role of Medium Access Protocols is to regulate the access of the sensor nodes to the shared wireless medium; this is, to the "air". However, we will first define some important metrics, which will help you identify how well a medium access protocol (MAC protocol) is performing.

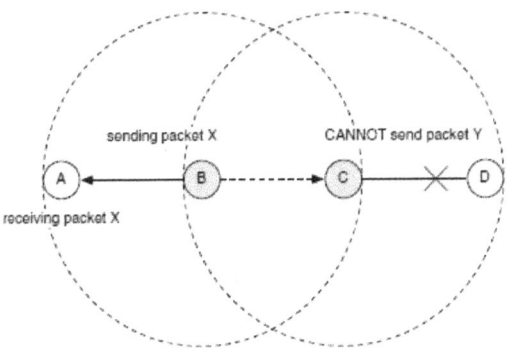

Figure 2.3: The exposed terminal problem in wireless communications.

Node B sends a packet to node A. At the same time, node C wants to send a packet to node D, and as the transmission ranges suggest, both transmissions are possible simultaneously. However, node C is prevented from sending, because it hears the transmission of B but does not know where A is.

Definition of Throughput: Throughput is defined as the number of bits or bytes transmitted successfully per time unit. Throughput is measured in bits per second.

Note that the throughput can be different in both directions-for sending and receiving. The throughput shows how much data you can push through a given medium whether cable, wireless, or even a sensor node. In terms of channel throughput, your interest should be in how many bits or bytes you can send over this channel so that all of them successfully arrive at the other side. When looking at the throughput of a sensor node, your interest should be in how many packets you can process and send out without issue. A MAC protocol typically attempts to maximize the throughput both at individual nodes and for the wireless medium in general. Furthermore, it tries to ensure some sort of fairness. This means that each node should have a fair chance to send out its packets. Another rather useful metric is the delay.

Definition of Delay: Delay is the amount of time between sending and receiving the packets.

Delays can be very short when measured internally between two hardware components and extremely large, e.g., when measured between terrestrial and satellite nodes. The delay provides you with information on how much time passes between creating a packet and its real arrival at the destination. A MAC protocol tries to minimize this time for all involved parties.

Design Criteria for Medium Access Protocols:

We are now ready to look at the design criteria for good MAC protocols. A MAC protocol has a hard problem to solve. It needs to maximize the throughput by minimizing the delay and energy spent. Additionally, for sensor networks, it needs to be able to handle switched off devices so nodes do not waste any valuable energy. Not surprisingly, the node spends much less energy in sleep mode than in active mode. This means you need to minimize the time in listening mode as much as possible. This also leads to the four main design criteria for medium access protocols.

- **Minimize collisions.** By avoiding packet collisions, the MAC protocol also avoids resending packets, which of course increases the throughput and decreases the energy spent. This task is rather difficult as the MAC protocol needs to orchestrate all nodes.

- **Minimize overhearing.** Overhearing is when a node receives a packet, which was not destined for it. Sometimes this can be useful, e.g., to see what is happening around. However, there are smarter ways to control who is receiving the packet and who is not (by using destination addresses in the header, for example). Thus, overhearing needs to be avoided because the overhearing node is supposed to trash packets which are not destined to it. Again, this is a hard task because the node needs to know when a packet is destined for it and when not and to go to sleep accordingly.

- **Minimize idle listening.** This problem is similar to overhearing in terms of energy waste because the node uses the same amount of energy in idle listening and receiving modes. It refers to the mode when the node is simply listening to the channel and nothing happens. This time needs to be minimized to save energy. Typically, going to sleep is not a problem but deciding when to wake up again is more challenging.

- **Minimize overhead.** Every packet and every bit, which does not carry application data (such as temperature or events), is considered an overhead. Even the destination and sender addresses are overhead. Each bit uses energy when sent or received and thus needs to be avoided, if not absolutely necessary. Completely avoiding the overhead is considered impossible with current technologies and is left to future generations to achieve (feel free to attack the problem).

Time Division Multiple Access:

The easiest way to organize communications in a network is by time. This is also called time division multiple access (TDMA). The basic principle is similar to "**divide and conquer**". Divide the time available across the nodes and give them full control over their slots. To achieve this, time is divided into rounds and rounds are divided into slots (Figure 2.7). Each node is then given sending control over one slot. The length of the slot depends on the technology used, clock precision of the sensor nodes, and on the expected length and number of the packets to be sent in one slot.

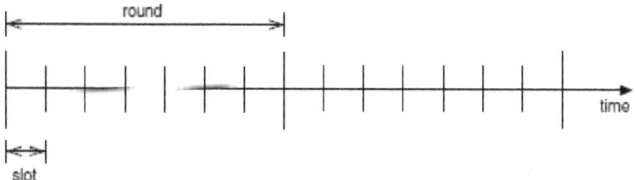

Figure 2.4: Rounds and slots in a TDMA schedule

The number of slots depends on the number of nodes in the network. The length of one slot and thus the length of a round depends on the technology used and on the expected length of the packets. It also depends on the precision of the sensors' internal clocks. Figure 2.4 presents the generalized TDMA algorithm. However, this is the algorithm for how TDMA works after the schedule has been set up. There are two possible ways to set up the schedule: centralized and distributed.

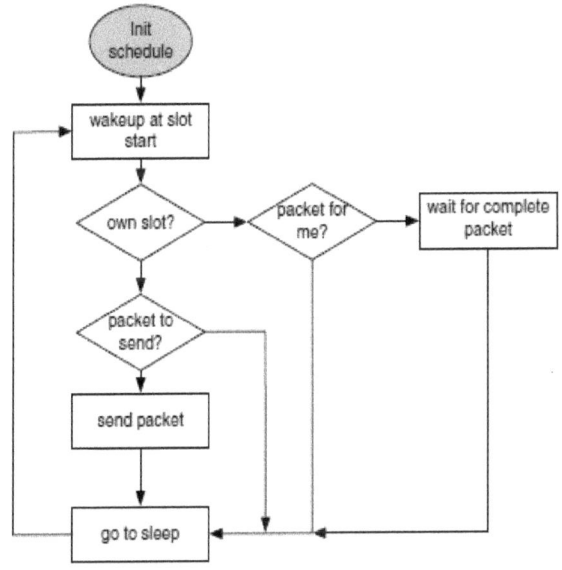

Figure 2.4: General TDMA MAC algorithm after schedule time.

Centralized TDMA: In centralized TDMA, the schedule is calculated offline and provided to the sensor nodes at startup. If nothing is known about the topology or the individual links between the nodes, the only option is to provide exactly N slots for N nodes and to give them control over the slots with their corresponding ID. Of course, if information about the topology, possible links and interferences is available, the task becomes more complex in computation, but the result can be more efficient. For example, non-interfering nodes can use the same slot and thus minimize the delay in the system.

The main disadvantage of centralized TDMA is its rigidness. Especially if the schedule takes into account the topology or the individual links between nodes, any slightest change in those will result in a complete mess in the schedule and harsh interference. On the other side, if the network is small and traffic low, centralized TDMA with N slots for N nodes is probably the simplest way to avoid interference and to allow for sleep.

Distributed TDMA: In the distributed variant of TDMA, the nodes attempt to find a good schedule by cooperation. Typically, the system starts with a predefined number and length of slots. The nodes start in the CSMA communication style, which means that they compete for the channel (that style is explored next). They first exchange neighbor information in terms of link quality. Next, they compete for the slots by trying to reserve them then release them again if interference occurs. When the system stabilizes, the network is ready to enter the final TDMA phase and to start working collision-free. Distributed TDMA is much more flexible and can also take into account the nodes' individual traffic needs. However, the initialization phase is quite complex and there is no guarantee that it will successfully converge. At the same time, once the schedule is found and no changes in the topology occur, communication is very efficient and fully collision-free. Even if changes occur, the protocol is able to detect them and to restart the initialization phase, if needed.

2.3 Wireless Links Introduction

The previous chapter explored wireless communications and their properties. It also discussed the problems associated with wireless communications, such as noise and interference and the practical problems arising from them, e.g., the hidden terminal problem or exposed terminal problem. At the level of MAC protocols, all we need is to understand when a node is allowed to access the shared wireless channel, without disrupting the work of others. The MAC protocol, however, does not guarantee that the transmission will be successful and also does not try to guarantee it.

In some cases, the MAC protocol uses acknowledgements to retransmit a packet. However, the MAC protocol cannot say how often it will need to retransmit the packet in order to succeed. This is the task of link layer protocols, which abstract the network as a set of links between individual nodes and attempt to characterize those links in terms of their quality or reliability. Link quality protocols pass this information to other protocols such as routing protocols. Moving away from technical details, the following summarizes what every user and developer of sensor networks needs to be most aware of.

2.4 Metrics of Wireless Links

- **Packet Reception Ratio or Rate (PRR):** This is the ratio between successfully delivered packets and all sent packets. PRR is measured in percentage and easily computable on any node without needing any special hardware.

- **Received Signal Strength Indicator (RSSI):** This is the signal strength, expressed in dBm (decibel per meter), for each received packet. This information is provided by the radio transceiver itself, but is often encoded, e.g., a node provides the results as a number, which later needs to be converted to dBm. The higher the RSSI, the better the signal. However, typical signal strengths are negative.

- **Link Quality Indicator (LQI):** This is a score given for each individual packet by the radio transceiver. It is part of the IEEE 802.15 standard, but its implementation is different for various vendors. In general, it is a positive number ranging from approximately 110 to 50, where higher values indicate better quality.

- **Signal to Noise Ratio (SNR):** This is the difference between the pure signal and the background noise, given in dB.

Wireless links are highly unpredictable and time-varying phenomena. The study of their properties could fill whole libraries and the details are highly scientific. The following is a summary of the most fundamental and relevant properties, which will help us manage and understand our sensor networks more efficiently.

2.5 Error Control

Errors happen all the time. However, you need to control these errors, recognize them, and even prevent them to enable efficient applications. Error control is a very important concept but often neglected by researchers. Besides some basic approaches, it is rare to see a real-world application with fully implemented error control. The main purpose of error control is to guarantee communication is error-free, in- sequence, duplicate-free, and loss-free. These properties are adopted from Internet-based communications, in which packets are parts of frames and their sequence is important. Sensor networks work in a slightly different way and some of the previous requirements are largely relaxed. For example, in sequence does not play a very important role because packets are self-sustainable and can easily be put back in sequence given their creation timestamps. Duplicates are not welcome, but also not damaging, as they can be also easily recognized and deleted at the sink. What remains are these two requirements:

- **Error-free.** Errors result either in faulty sensor data, which is harmful to the application, or in resending of packets, which is harmful to the network lifetime and throughput.

- **Loss-free.** Losses are harmful to the application and might render the whole network useless.

In terms of sensor networks, the following discussion focuses on these two types of problems. In general, backward error control and forward error control are the two types of error control schemes that are used.

Backward Error Control:

Backward error control refers to the fact that you should not try to prevent errors but only discover them. Once we discover a problem with a particular packet received, we should send the so-called Automatic Repeat Request (ARQ) to the sender. How to discover the error? The simplest way is to use a Cyclic Redundancy Check (CRC). This is a function computed over the contents of the packet and attached to it. For example, you could compute the sum of all bits in the packets, often referred to as the checksum. Figure 2.5 provides an example of a packet with a computed checksum and an error discovery on the receiver's side. In this case, we have used the simplest CRC checksum, the parity bit. The parity bit computes whether the number of 1s in the packet is even or odd, marking an even number with 0 and an odd number with 1. This is the same as an AND computation over the bits.

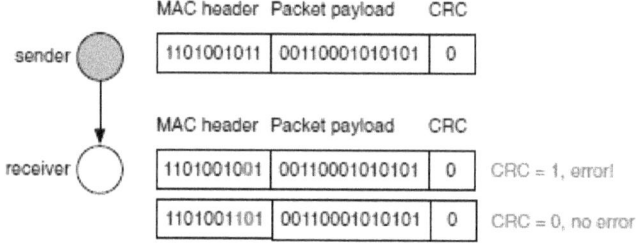

Figure 2.5: Computation of CRC and error detection of a sample

Of course, the parity bit is simple and does not detect all errors. For example, in the first shown packet for the receiver in Figure 2.5, there is only one error, transforming a 0 into 1. This error is correctly detected by the parity bit. However, the second packet has two errors, which neutralize each other. The computed parity bit corresponds to the received one, thus no error is detected. For this reason, more complex CRC code computations with more bits exist.

Forward Error Control:

Forward Error Control (FEC) attempts to prevent errors instead of only detecting them. This is reasonable, because this is due to the overhead involved in sending a single packet. Recall MAC protocols from previous chapter. They involve a lot of administrative tasks before you can actually send the packet out, like the long preambles of BMAC or the CTS/RTS handshaking of CSMA/CA. Thus, it is a good idea to avoid resending a packet if you can prevent errors by adding more information into the first packet. The simplest FEC technique is to repeat the packet payload several times in the same packet, called repeating code and shown in Figure 2.6. Here, the payload can be reconstructed if not too many errors happened to one of the repeated bytes.

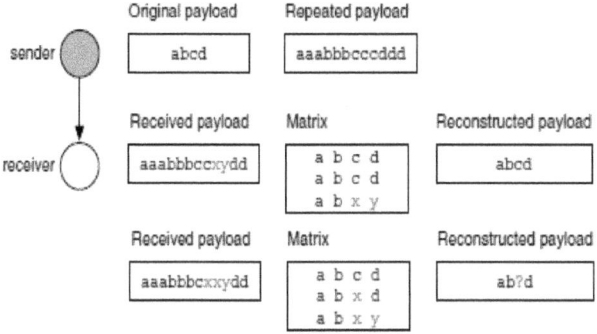

Figure 2.6: FEC Scheme

2.6 Naming and Addressing

One question to resolve before moving on to link quality protocols is how to address individual nodes. Fundamental differences are made between naming and addressing. While naming resembles more human names, which are usually kept forever, addressing is more like your current home address.

Naming:

Naming gives some abstract identification to individual nodes. No real restrictions apply, but typically it is practical to use unique names in the same network and to represent them in an efficient way. The most common way to name nodes in a sensor network is to give them IDs, which are represented by integers and are unique. It is very important to understand that a name does not reveal anything about the position of the node. However, it can reveal something about the role of the node (e.g., sink or gateway) or the type of data it can deliver. An extreme case of the latter is the so-called data naming, where nodes with the same type of sensors share also data naming the same name. For example, all nodes with temperature sensors have ID 1 and all nodes with light sensors have ID 2. This is sometimes useful, particularly when the application does not care where the data comes from, but only what kind of data is delivered.

Addressing:

Different from naming, addressing reveals more about the position of the node than its role or abilities. A typical example of addressing is IP addresses, which clearly identify the position of individual nodes given their addresses. In sensor networks, IP addresses are rarely used besides maybe for the sink or gateway of the network. More often, the address is the direct geographical or relative position of the node.

Assignment of Addresses and Names:

For small networks, it might seem trivial to assign unique names or addresses to nodes. You could just take IDs 1, 2, 3, for names and the real GPS positions of the nodes as their addresses. However, both become important with larger networks or with randomly deployed networks. Protocols and algorithms for automatic name and address assignment are out of scope for this book, but the following should be kept in mind:

- **Size and length:** You should plan for scalability and not rely on the fact that the network will always remain small. The transition from IPv4 to IPv6 addresses is an example of a problem this thinking could create later on. IPv4 uses 32-bit addresses, which corresponds to a maximum of 4.3 billion addresses. With the explosion of the Internet of Things, this number became too small and IPv6 addresses were introduced with 128 bits, capable of representing an amazing 3.4×1038 addresses or 40,000 addresses for each atom on Earth. With sensor networks, you should not go that far but reserving several bytes might be better than a single one.

- **Storage:** The next problem is where to store the address and the name. If stored in the normal volatile memory, the node will loose the information every time it reboots.

- **Positioning and repositioning:** Another critical point is how to identify the address of a node and how to update it if the node moves away. Chapter 8 addresses this problem.

- **Uniqueness:** Even if some applications permit reuse of names, it is typically important to have unique names in the same network. For example, this is easy to achieve when new nodes are simply given incremental IDs. This is also the reason why a large count of unique IDs is important. Another very common option is to give the nodes their own fabrication numbers or MAC addresses as IDs. These are, however, typically very long.

Using Names and Addresses:

- Names are typically too abstract whereas addresses are too exact. Thus, in sensor networks a combination is very often used, where the name (ID) of the node points to the identity of the node and some of its properties or abilities and the address represents its current position (relative or absolute). For some purposes, such as link quality prediction (which are explored next) only the name is really relevant. For other protocols, such as routing, the addresses are more relevant and the ID becomes less important.

2.7 Topology Control

Until now, discussion has focused on how to measure existing wireless links as well as their properties and quality. However, you can also control these properties, for example, by increasing and decreasing the transmission power. The higher the transmission power, the more energy you use and the better the link becomes. The lower the transmission power, the more restricted the transmission area is not necessarily a disadvantage. Very dense networks often suffer from overloading and too many neighbors. Allowing each node to use its own transmission power poses several communication challenges:

- **Asymmetric links** become the rule, as two communication nodes are allowed to use different transmission powers.

- **Directional links** become possible, but are rather undesired. While in theory nodes can save energy by using only the minimally required transmission power to send data traffic and avoid any acknowledgements or control beacons, this often renders the whole system fragile and unreliable. If something goes wrong, the sending node has no chance of discovering the problem.

- **Discovery of neighbors** becomes challenging because they could have too low transmission power to answer a query.

Given the proceeding issues, it is rather rare to see a topology control protocol that allows for a completely free choice of transmission power and especially selfish decisions. Typically, either unified transmission power is selected for all nodes or a transmission power is selected so that communication neighborhoods can successfully talk to each other. Apart from these options, there are two possibilities of how to agree on the transmission power: centralized or distributed.

Centralized Topology Control:

In centralized protocols, one of the nodes takes a leading role in the process and disseminates its decision to all other nodes, which have to obey. This is typically a more important or powerful node, such as the base station or the sink. In order for this approach to work, the nodes have to send some communication information to the sink to analyze and take decisions. There are two major options:

- The nodes try out with their current transmission power and store the received RSSI and LQI values. They send them to the sink and the sink analyzes them to decide whether to lower the transmission power and by how much.

- The sink first sends a command to all nodes to change their transmission power and to report on the new available neighbors.

Typically, a combination of both approaches is used, in one or another order. However, the major problem comes from the fact that if the sink commands the node to go to a low transmission power, they might not be able to report anything. They are only able to receive a new command later on (the transmission power change affects only the sending process, not the receiving one). Of course, the sink could decide also for different transmission powers for individual nodes, but this is typically avoided because of too high complexity. Centralized control is usually simpler, but it also requires significant communication overhead and does not react well in case of changes.

Distributed Topology Control:

In distributed control, the optimal transmission power is adjusted at individual nodes without having somebody to control or oversee the process. Nodes try out different transmission powers and evaluate their neighborhood.

The greatest challenge here is to synchronize more or less the process at neighboring nodes so that neighbors can understand each other. For example, node 1 and node 2 try to find their minimal transmission powers so that they can still communicate to each other. Remember that they could be different because of asymmetric links. Node 1 incrementally decreases its transmission power and every time it does so, it sends a request to node 2 to check whether it can still reach it. If node 2 does not play currently with its transmission power, this approach would work well and node 1 will soon discover the transmission power at which it can easily reach node 2.

However, if node 2 is also playing around while replying to node 1's requests, the situation becomes tricky. If node 1 does not receive the reply, what is the reason? That its own transmission power is too low and its request was not received at all at node 2? Or because node 2's transmission power is too low? The solution to this problem is to allow only for maximum transmission power when sending replies.

Thus, the nodes can play simultaneously with their transmission powers but when they answer requests from other nodes, they have to switch to the maximum. Distributed control is more flexible, situation-aware, and precise than centralized control.

•●•

UNIT - 3

Wireless Standards and Protocols

3.1 IEEE 802.15.4 Low Rate WPAN

IEEE 802.15.4 is the proposed standard for low-rate wireless personal area networks (LR- WPAN's) with focus on enabling WSNs. IEEE 802.15.4 focuses on low cost of deployment, low complexity, and low power consumption; it is designed for wireless sensor applications that require short range communication to maximize battery life. WSNs applications using IEEE 802.15.4 include residential, industrial, and environment monitoring, control and automation. IEEE 802.15.4 devices are designed to follow the physical and data-link layer protocols. IEEE 802.15.4 compared with 802 wireless standards is illustrated in Table 3.1.

Parameters	802.11b WLAN	802.15.1 WPAN	802.15.4LR-WPAN
Range	~ 100 m	~ 10-100 m	10 m
Raw Data rate	11 Mbps	1 Mbps	<= 0.25 Mbps
Power consumption	Medium	Low	Ultra-Low

Table 3.1: IEEE 802.15.4 compared with 802 wireless standards

The IEEE 802.15.4 standard allows the formation of the star and peer-to-peer topology for communication between network devices (Figure. 3.1).

- **Star Topology:** In the star topology, the communication is performed between network devices and a single central controller, called the PAN coordinator. A network device is either the initiation point or the termination point for network communications. The PAN coordinator is in charge of managing all the star PAN functionality.

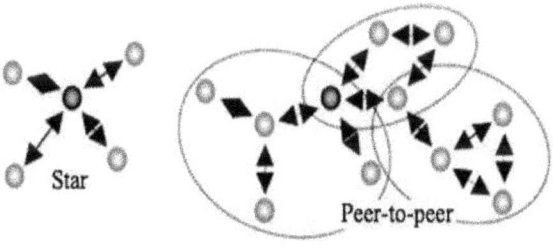

Figure 3.1: Star and peer-to-peer topology organized as cluster network

- **Peer-to-Peer Topology**: In the peer-to-peer topology, every network device can communicate with any other within its range. This topology also contains a PAN coordinator, which acts as the root of the network. Peer-to-peer topology allows more complex network formations to be implemented; e.g., ad hoc and self-configuring networks. The routing mechanisms required for multi-hopping are part of the Network layer and are therefore, not in the scope of IEEE 802.15.4.

3.2 ZigBee

The ZigBee standard was publicly available as of June 2005 (ZigBee Alliance 2013), it defines the higher layer communication protocols built on the IEEE 802.15.4 standards for LR-PANs. ZigBee got its name from the way bees zig and zag while tracking between flowers and relaying information to other bees about where to find resources. ZigBee is a simple, low cost, and low power wireless communication technology used in embedded applications. ZigBee devices use Very little power and can operate on a cell battery for many years. ZigBee has been introduced by IEEE with IEEE 802.15.4 standard and the ZigBee Alliance to provide the first general standard for such applications. ZigBee is built on the robust radio (PHY) and medium access control (MAC) communication layers defined by the IEEE 802.15.4 standard for LR-WPANs. On the higher layer, ZigBee defines mesh, star and cluster tree network topologies with data security features and interoperable applications.

Table 1.2 compares ZigBee with wireless standards that address mid to high data rates for voice, PC LANs, video, etc. However, ZigBee meets the unique needs of sensors and control devices, typically, low bandwidth, low latency and very low energy consumption for long battery lives and for large device arrays. ZigBee is simpler than Bluetooth, it has a lower data rate and spends most of its time sleepy. It is accepted that standards such as Bluetooth and WLAN are not suited for low power applications, due to their high node costs as well as complex and power demanding RF- ICs and protocols.

Introduction to Wireless Sensor Networks

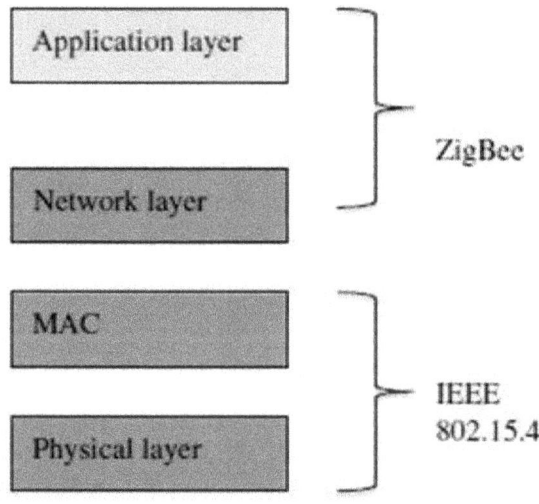

Figure 3.2: ZigBee over IEEE 802.15.4 buildup

	Blueto oth	UWB	Zigbee	Wi-Fi
IEEE Specification	802.15.1	802.15.3a	802.15.4	802.11a/b/g
ISM frequency Band	2.4 ghz	3.1-10.6 ghz	868/915 mhz, 2.4 ghz	24 Giltz, 5 ghz
Application	Wireless connectivity between devices such as phone, PDA,	Real-time video and music, multimedia wireless networ, WPAN	Industrial control and monitoring. Sensor networks, building automat	Wireless LAN connectivity, broadband Internet

	laptop, headsets		ion, home control and automation, toys, games	
Max signal rate	1 Mbps	110 Mbps	250 Mbps	54Mbps
Nominal range	10 m	10 m	10-100	100 m
Transmission power	0-10 dbm	-41.3 dbm/mhz	(-25)-0 dbm	15-20 dbm
Channel bandwidth	1 Mbps	500-7.5 ghz	0.3/0.6; 2 mhz	22 mhz
Modulation type	GFSK	BPSK, QPSK	BPSK (+ASK) O-QPSK	BPSK, QPSK COFDM, CCK, M-QAM
Basic cell	Piconet	Piconet	Star	BSS
Extension of the basic cell	Scatternet	Peer-to-peer	Cluster tree, Mesh	ESS
Max number of cell nodes	8 active devices, 255 in park mode	8	>65,000	Unlimited in ad hoc networks (IBSS), up to 2007 devices in infrastructure networks

Encryption	EO stream cipher	AES block cipher (CTR (counter mode)	AES block cipher (CTR (counter mode)	RC4 stream cipher (WEP), AES block cipher
Authentication	Shared secret	CBC-MAC (CCM)	CBC-MAC (ext. Of CCM)	WPA2 (802.11i)
Data protection	16-bit CRC	32-bit CRC	16-bit CRC	32-bit CRC

Table 3.2 Different Wireless Technologies Comparison

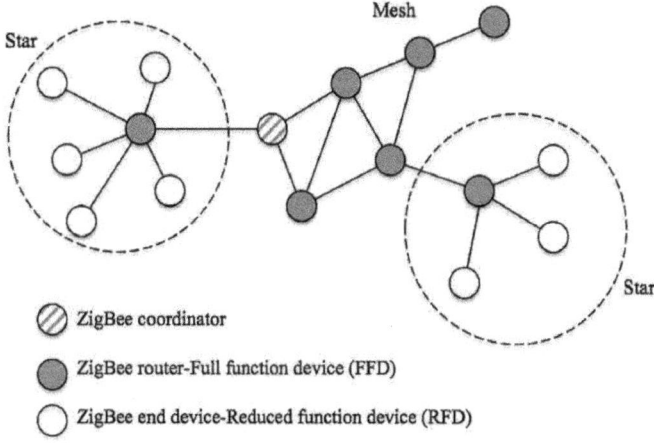

Figure 3.3: ZigBee network model

- ZigBee coordinator, it initiates network formation, stores information, and can bridge networks together.

- ZigBee routers, they link groups of devices together and provide multi-hop communication across devices.

- ZigBee end device, it consists of the sensors, actuators, and controllers that collects data and communicates only with the router or the coordinator.

3.2.1 Wireless HART

Wireless HART was released in September 2007 (Kim et al. 2008; Song et al.2008). The Wireless HART standard provides a wireless network communication protocol for process measurement and control applications, it is based on IEEE 802.15.4 for low power 2.4 GHz operation. Wireless HART is compatible with all existing devices, tools, and systems, it is reliable, secure, and energy efficient, and supports mesh networking, channel hopping, and time-synchronized messaging. Network communication is secure with encryption, verification, authentication, and key management. Power management options enable the wireless devices to be more energy efficient. Wireless HART is designed to support mesh, star, and combined network topologies. As shown in Fig. 1.4, Wireless HART network consists of wireless field devices, gateways, process automation controller, host applications, and network manager:

- Wireless field devices are connected to process or plant equipment.

- Gateways enable the communication between the wireless field devices and the host applications.

- Handheld which is a portable Wireless HART enabled computer used to configure devices, run diagnostics, and perform calibrations.

- The network manager configures the network and schedule communication between devices, it also manages the routing and network traffic. The network manager can be integrated into the gateway, host application, or process automation controller.

Figure 3.4 illustrates the architecture of the Wireless HART protocol stack in accordance with the OSI 7-layer communication model.

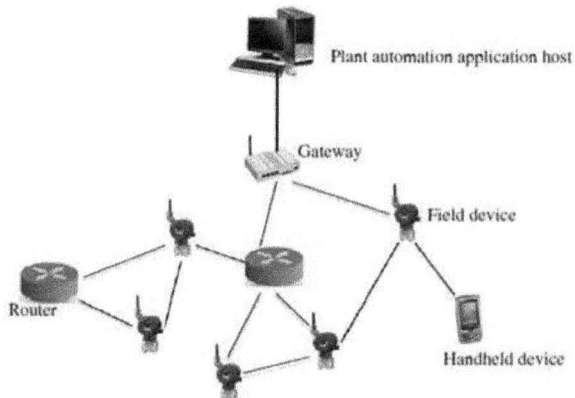

Figure 3.4: Wireless HART mesh networking

3.2.2 ISA100.11a

ISA100.11a, was officially approved in September 2009 by ISA Standards and Practices Board. It is the first standard of ISA100 family with foundations for process automation, and provisions for secure, reliable, low data rate wireless monitoring. Specifications for the OSI layer, security, and system management are comprised

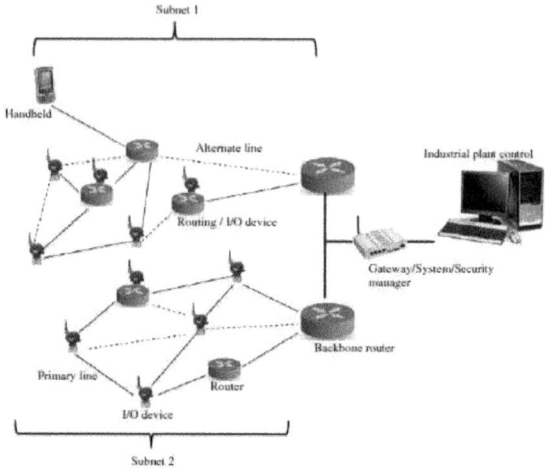

Figure. 3.5 ISA 100.11a mesh networking

ISA100.11a defines for devices different role profiles that represent various functions and capabilities, such as I/O devices, routers, provisioning devices, backbone routers, gateway, system manager, security manager. Each device may resume more than a role, while its capabilities are reported to the system manager upon joining the network

- I/O device (sensor and actuator), provides or/and consumes data, which is the basic goal of the network.

- Handheld which is a portable computer used to configure devices, run diagnostics, and perform calibrations.

- Router is a job accorded to devices responsible for routing data packet from source to destination and propagating clock information. A router role also enables a device to act as a proxy that permits new devices to join the network.

- Device with provisioning role, for pre-configuring devices with necessary information to join a specific network.

- Backbone router, routes data packets from one subnet connected to the backbone network to a destination (e.g., another subnet connected to the backbone).

The backbone router is implemented with both ISA100.11a wireless network interface and backbone interface.

- Gateway, acts as an interface between ISA100.11a field network and the host applications in the control system.

- System manager, is the administrator of the ISA100.11a wireless network. It monitors the network and is in charge of system management, device management, network run-time control, and communication configuration (resource scheduling), as well as time related services.

- Security manager, provides security services based on policies specified by this standard.

3.3. 6LoWPAN

IPv6-based low power wireless personal area networks (6LoWPAN) enables IPv6 packets communication over an IEEE 802.15.4 based network (Mulligan 2007; Montenegro et al. 2007; Shelby and Bormann 2011). Low power devices can communicate directly with IP devices using IP-based protocols. Utilizing 6LoWPAN, low power devices have all the benefits of IP communication and management. 6LoWPAN standard provides an adaptation layer, new packet format, and address management. Because IPv6 packet sizes are much larger than the frame size of IEEE 802.15.4, the adaptation layer is used. The adaptation layer accomplishes header compression, which creates smaller packets fitting into the IEEE 802.15.4 frame size. Address management mechanism handles the forming of device addresses for communication. 6LoWPAN is designed for applications with low data rate devices that require Internet communication The Wireless Embedded Internet is created by connecting networks of wireless embedded devices, each network is a stub on the Internet. A stub network is a network where IP packets are sent from or destined to, but which does not act as a transit to other networks. The 6LoWPAN architecture is made up of low-power wireless area networks (LoWPANs), which are IPv6 stub networks. The overall 6LoWPAN architecture is presented in Figure 3.6. A LoWPAN is the collection of 6LoWPAN nodes which share a common IPv6 address prefix (the first 64 bits of an IPv6 address), meaning that regardless of where a node is in a LoWPAN its IPv6 address remains the same. Three different kinds of LoWPANs have been defined,

- Simple LoWPAN, connected through one LoWPAN Edge Router to another IP network. A backhaul link (point-to-point, e.g., GPRS) is shown in the figure, but it could also be a backbone link (shared).

- Extended LoWPAN, that encompasses the LoWPANs of multiple edge routers via a backbone link (e.g.,

Ethernet) interconnecting them. Edge routers share the same IPv6 prefix and the common backbone link.

- Ad hoc LoWPAN, that is not connected to the Internet, but instead operates without an infrastructure. A LoWPAN consists of one or more edge routers along with nodes, which may function as host or router. The network interfaces of the nodes share the same IPv6

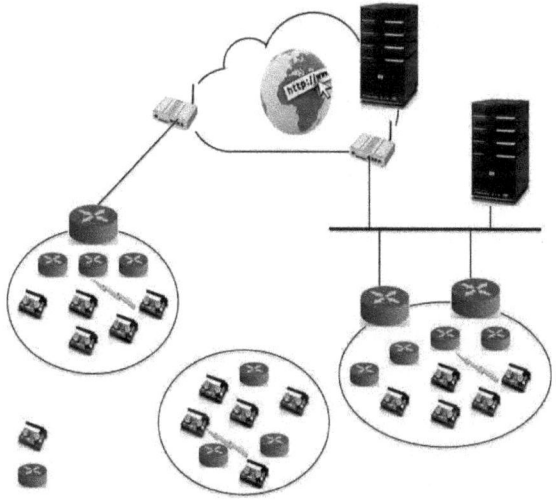

Figure 3.6 Ad hoc LoWPAN

Figure 3.6 shows the 6LoWPAN architecture (based on Shelby and Bormann 2011) prefix distributed by the edge router and routers throughout the LoWPAN. Each node is identified by a unique IPv6 address, and is capable of sending and receiving IPv6 packets. In order to facilitate efficient network operation, nodes register with an edge router. LoWPAN nodes may participate in more than one LoWPAN at the same time (called multi- homing), and fault-tolerance can be achieved between edge routers.

Nodes are free to move throughout the LoWPAN, between edge routers, and even between LoWPANs. Topology change may also be caused by wireless channel conditions, without physical movement. LoWPANs are connected to other IP networks through edge routers, as seen in Figure. The edge router plays an important role as it routes traffic in and out of the LoWPAN, while handling 6LoWPAN compression and Neighbor Discovery for the LoWPAN. If the LoWPAN is to be connected to an IPv4 network, the edge router will also handle IPv4 interconnectivity. Edge routers have management features tied into overall IT management solutions. Multiple edge routers can be supported in the same LoWPAN if they share a common backbone link.

3.3.1 IEEE 802.15.3

IEEE 802.15.3 as proposed in 2003 is a MAC and PHY standard for high-rate WPANs (11−55 Mbps) (Tseng et al. 2003; IEEE 2013). IEEE 802.15.3a was an attempt to provide a higher speed UWB PHY enhancement amendment to IEEE 802.15.3 for applications that involve imaging and multimedia. But, the proposed PHY standard was withdrawn in 2006 as the members of the task group were not able to come to an agreement choosing between two technology proposals, multi-band orthogonal frequency division multiplexing (MB-OFDM) and direct sequence UWB (DS-UWB), backed by two different industry alliances.

The IEEE 802.15.3b-2005 amendment was released on May 5, 2006. It enhanced 802.15.3 to improve implementation and interoperability of the MAC. This includes minor optimizations while preserving backward compatibility. In addition, this amendment corrected errors, clarified ambiguities, and added editorial clarifications. IEEE 802.15.3c- 2009 was published on September 11, 2009. The IEEE 802.15.3 Task Group 3c (TG3c) was formed in March 2005. TG3c developed a millimeter-wave-based alternative physical layer (PHY) for the existing 802.15.3 WPAN Standard 802.15.3-2003.

This millimeter-wave WPAN operates in clear band including 57–64 GHz unlicensed band defined by FCC 47 CFR 15.255. The mm WPAN permits high coexistence (close physical spacing) with all other microwave systems in the 802.15 family of WPANs. In addition, the mm WPAN allows very high data rate over 2 Gbps applications such as high-speed internet access, streaming content download (video on demand, HDTV, home theater, etc.), real-time streaming and wireless data bus for cable replacement. Also, there are optional data rates in excess of 3 Gbps.

At the MAC layer, WPAN high-rate technology (802.15.3) is based on centralized signaling and peer-to-peer traffic structure; the nodes are classified as PicoNet Coordinators (PNC) and Devices (DEV). A PNC assigns guaranteed time slots to all nodes for communication. More precisely, there is a period for contention, followed by a contention free period, which contains guaranteed time slots being allocated by the PNC as shown in Figure 3.7.

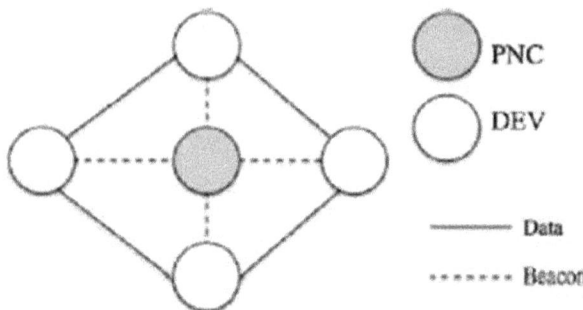

Figure 3.7 IEEE 802.15.3 MAC Network structure

3.4. Wibree, BLE

Released in 2006 by Nokia, it is a wireless communication technology designed for low power consumption, short-range communication, and low-cost devices; it is called Baby-Bluetooth, and renamed Bluetooth Low Energy (BLE) technology (Peiet al. 2008). Wibree allows the communication between small battery-powered devices and Bluetooth devices. Small battery powered devices include watches, wireless keyboard, and sports sensors which connect to host devices such as personal computer or cellular phones. This standard operates on 2.4 GHz and has a data rate of 1 Mbps, with 5–10 m as a linking distance between the devices. Wibree may be deployed on a stand-alone chip or on a dual-mode chip along with conventional Bluetooth, it works with Bluetooth to make devices smaller and more energy-efficient. Bluetooth-Wibree utilizes the existing Bluetooth RF and enables ultra-low power consumption. A key point must be taken into consideration, BLE is incompatible with standard Bluetooth, BLE devices do not interoperate with classical Bluetooth products.

However, implementing a dual-mode device could achieve such interoperability. A dual-mode device is an integrated circuit that includes both a standard Bluetooth radio and a BLE radio, each mode operates separately, not at the same time, though they can share an antenna. Several vendors offer dual mode chips, such as Broadcom, CSR, EM Microelectronics, Nordic Semiconductor, and Texas Instruments. Complete modules also are available from connect Blue (Frenzel 2012). Table 3.3 compares Bluetooth and Wibree.

	Bluetooth	Wibree
Band (GHz)	2.4	2.4
Antenna/HW	Shared	
Power (mW)	100	10
Target battery life	Days—months	1–2 years
Peak current consumption (mA)	<30	<15
Range (m)	10–30	10
Data rate (Mbps)	1–3	1
Application throughput (Mbps)	0.7–2.1	0.27
Active slaves	7	Unlimited
Component cost	$3	Bluetooth + 20¢
Network topologies	Point to point, scatternet	Point to point, star
Security	56–128 bit encryption	128-bit AES
Time to wake and transmit (ms)	100+	<6

Acronyms
AES (advanced encryption standard)

Table 3.3: Comparison of Bluetooth and Wibree Technologies

3.4.1 Z-Wave

Z-Wave, a proprietary technology developed by Zensys A/S of Denmark, is focusing exclusively on the residential market (Reinisch et al. 2007; Z-Wave Alliance 2012). The two wireless networking standards, Z-Wave and ZigBee, are competing to become the standard for automated home control. ZigBee, an IEEE802.15.4 based standard proposed by a large group of worldwide manufacturers represented by the ZigBee Alliance, has a broader focus that includes both home and commercial control systems (Fig. 1.19). Z-Wave uses a two-way RF system that operates in the 908 MHz ISM bands (868 MHz in Europe and 908 MHz in the United States). Z-Wave allows transmission at 9.6 and 40 Kbps data rates using binary frequency shift keying (BFSK) modulation.

The recent Z-Wave 400 series single chip supports the 2.4 GHz band and offers bit rates up to 200 Kbps. In contrast to other wireless networking technologies such as Bluetooth and wireless LAN, Z-Wave features lower power consumption and lower data rates. With very short transmit times and efficient design, Z-Wave nodes can easily be powered from a battery with long lifetime. Applications like residential lighting control take no more than 250 ms. Z-Wave relies on the fact that its targeted residential applications require the transmission of small amounts of data, and therefore it uses a data rate of just 9.6 Kbps.

The second-generation Z- Wave chipset, the ZW0201, is used as a mixed signal chip, integrating an RF transceiver, Z- Wave protocol storage and handling, and capacity for application storage and handling. The ZW0201 as the core of Z-Wave features a low standby current of 0.1 u A. This current rises to 25 mA on transmission, but as the protocol has been designed to keep transmit and receive time to an absolute minimum it is possible to run a node from a battery. The device includes an 8-bit CPU core running at 8 MHz with up to 32 Kbits of flash memory; it has enough capacity to handle both an application as well as the wireless communication protocol. On top of the link layer, a source-routing protocol gives designers the ability to setup a Z-Wave mesh network. Based on the network topology data in the initiator's memory, Z-Wave's source-routing protocol allows the initiator to generate a complete route from the initiator to the destination. Z-Wave defines two types of devices, controllers and slaves. Controller's poll or send commands to the slaves, which reply to the controllers or execute the commands.

The Z-Wave routing layer performs routing based on a source routing approach. When a controller transmits a packet, it includes the path to be followed in the packet. A packet can be transmitted over up to four hops, which is sufficient in a residential scenario and hard-limits the source routing packet overhead. A controller maintains a table that represents the full topology of the network. A portable controller (e.g., a remote control) tries first to reach the destination via direct transmission; if that option fails, the controller estimates its location and calculates the best route to the destination.

Slaves may act as routers; routing slaves store static routes (typically toward controllers) and are allowed to send messages to other nodes without being requested to do so. Slaves are suitable to be monitoring sensors where the delay contributed by polling is acceptable, as well as for actuators that perform actions in response to activation commands. Routing slaves are used for time critical and no solicited transmission applications such as alarm activation.

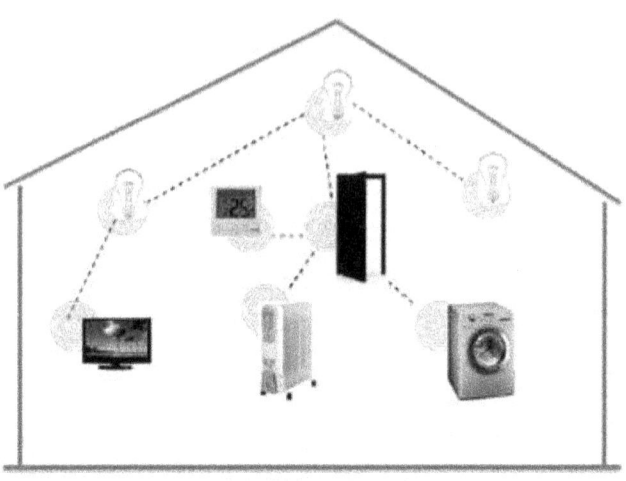

Figure 3.8 Z-Wave WSN Home Appliances Control

3.4.2 ANT

ANT is a proprietary technology featuring a wireless communication protocol stack thought for ultra-low power networking applications (ANT 2013). It is designed to run using low cost, low power micro-controllers and transceivers operating in the 2.4 GHz ISM band. The ANT WSN protocol has been engineered for simplicity and efficiency, resulting in ultra-low power consumption, maximized battery life, a minimal burden on system resources, simpler network designs and lower implementation costs. ANT also features low latency, ability to trade-off data rate against power consumption, support for broadcast, burst and acknowledged transactions up

to a net data rate of 20 Kbps (ANTs over the air data rate is 1 Mbps). Different topologies could be established, peer-to-peer, star, tree and other types of mesh network.

ANT nodes are capable of acting as slaves or masters within a network and swapping roles at any time. This means that the nodes can act as transmitters, receivers or transceivers to route traffic to other nodes. ANT is a good protocol for practical networks because of this inherent ability to support ad hoc interconnection of tens or hundreds of nodes. ANT allows a system to spend most of its time in an ultra-low power sleep mode, wake up quickly, transmit for the shortest possible time and quickly return back to an ultra-low power sleep mode.

This implies that ANT is one of the energy-efficient available technologies. While Bluetooth is designed for rapid file transfer between devices in a PAN, its average power consumption is 10 times greater with respect to ANT and the hardware costs are 90 % higher. With respect to IEEE 802.15.4, ANT presents a larger data rate of 1 Mbpsec and is relatively less complex. However, being a proprietary technology, ANT lacks interoperability. ANT+ is a relatively recent addition to ANT. This software function provides interoperability in a managed network; it facilitates the collection, automatic transfer, and tracking of sensor data for monitoring all involved nodes and devices.

But what is a managed network? It is a type of communication network that is built, operated, secured and managed by a third-party service provider, it is an outsourced network that provides some or all the network solutions required by an organization. SensRcore, an extra ANT feature, is a development system that helps developers create low-power sensor networks. ANT transceiver chips are available from Nordic

Semiconductor and Texas Instruments. A number of similarities exist between ANT and BLE (Sect. 8.7), but their differences are stark. Both are good choices for very low-power applications (Table 1.9), but:

- ANT has the simplest protocol with minimum overhead, and it supports more different types of network topologies.

- BLE is a star-only format, while ANT supports all types including mesh. More vendors offer Bluetooth chips and modules versus ANT, though.

3.4.3 INSTEON

INSTEON is a solution developed for home automation by SmartL abs and promoted by the INSTEON Alliance (INSTEON 2013). One of the distinctive features of INSTEON is the fact that it defines a mesh topology composed of RF and power line links. Devices can be RF-only or power-line-only, or can support both types of communication. INSTEON RF signals use frequency shift keying (FSK) modulation at the 904 MHz center frequency, with a raw data rate of 38.4 Kbps.

INSTEON networking has several features.

- INSTEON devices are peers, which means that any of them can play the role of sender, receiver, or relayer.

- Communication between devices that are not within the same range is achieved by means of a multi-hop approach that differs in many aspects from traditional techniques. All devices retransmit the messages they receive, unless they are the destination of the messages. The maximum number of hops for each message is limited to four (as in Z-Wave). The multi-hop transmission is performed using a time slot synchronization scheme, by which transmissions are permitted in certain time slots, and devices within the same range do not transmit different messages at the same time. These time slots are defined by a number of power line zero crossings.

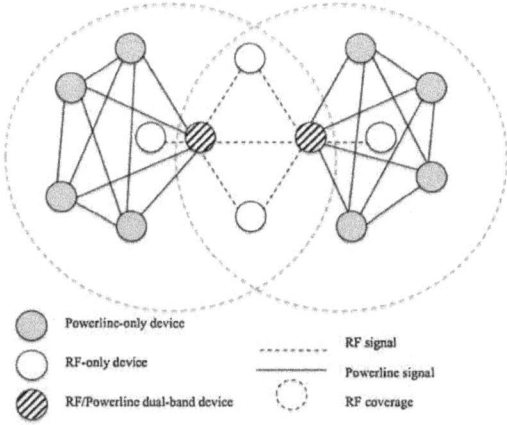

Figure 3.9 INSTEON Network Structure

- RF devices that are not attached to the power line can transmit asynchronously, but RF devices attached to the power line will retransmit the related messages synchronously.

- In contrast to classical collision avoidance mechanisms, devices within the same range are allowed to transmit the same message simultaneously. This approach, which is called simulcast, relies on the very low probability of multiple simultaneous signals being cancelled at the receiver.

3.4.4 Wavenis

Wavenis is a wireless protocol stack developed by Coronis Systems for control and monitoring applications in several environments, including home and building automation (Gomez and Paradells 2010). Wavenis is currently being promoted and managed by the Wavenis Open Standard Alliance (Wavenis-OSA). It defines the functionality of physical, link, and network layers. Wavenis services can be accessed from upper layers through an application programming interface (API). Wavenis operates mainly in the 433 MHz, 868 MHz, and 915 MHz bands, which are ISM bands in Asia, Europe, and the United States.

Some products also operate in the 2.4 GHz band. The minimum and maximum data rates offered by Wavenis are 4.8 Kbps and 100 Kbps, respectively, with 19.2 Kbps being the typical value. Data are modulated using Gaussian FSK (GFSK). Fast frequency-hopping spread spectrum (FHSS) is used over 50 kHz bandwidth channels.

The Wavenis MAC sub layer offers synchronized and non-synchronized schemes:

- In a synchronized network, nodes are provided with a mixed CSMA/TDMA mechanism for transmitting in response to a broadcast or multicast message. In such a case, a node allocates atime slot that is pseudo-randomly calculated, based on its address. Before transmission in that slot, the node performs carrier sense (CS). If the channel is busy, the node computes a new time slot for the transmission.

- For non-synchronized networks, in applications where reliability is a critical requirement (alarms, security, etc.), CSMA/CA is used. The Wavenis logical link control (LLC) sub layer manages flow and error control by offering per-frame or per-window ACKs. Wavenis defines only one type of device. The Wavenis network layer specifies a four-level virtual hierarchical tree. The root of the tree may play the role of a data collection sink or a gateway. A device that joins a Wavenis network intends to find an adequate parent, for this purpose, the new device broadcasts a request for a device of a certain level and a sufficient quality of service (QoS) value. The QoS value is obtained by taking into consideration parameters such as received signal strength indicator (RSSI) measurements, battery energy, and the number of devices that are already attached to this device.

3.5 Protocol Stack of WSNs

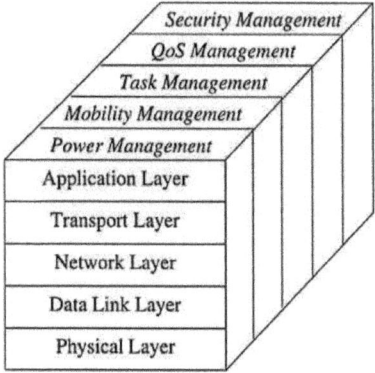

Figure 3.10 Protocol stack of WSNs (Wang and Balasingham 2010)

Physical Layer:

In many wireless sensor networks, the number and location of nodes make recharging or replacing the batteries infeasible. For this reason, energy consumption is a universal design issue for wireless sensor networks. Much work has been done to minimize energy dissipation at all levels of system design, from the hardware to the protocols to the algorithms. Hence to the network, it is important to appropriately set parameters of the protocols in the network stack. At the physical layer, the parameters open to the network designer include, modulation scheme, transmit power and hop distance. The optimal values of these parameters will depend on the channel model. When a wireless transmission is received, it can be decoded with a certain probability of error, based on the ratio of the signal power to the noise power of the channel (i.e., the SNR). As the energy used in transmission increases, the probability of error goes down, and thus the number of retransmissions goes down. Thus there exists an optimal tradeoff between the expected number of retransmissions and the transmit power to minimize the total energy dissipated to receive the data (Holland et al. 2011). At the physical layer, there are two main components that contribute to energy loss in a wireless transmission, the loss due to the channel and the fixed energy cost to run the transmission and reception circuitry.

Data Link Layer:

The responsibilities of the data link layer are the multiplexing of data streams, data frame detection, medium access (MAC) and error control. A wireless sensor network must have a specialized MAC protocol to address the issues of power conservation and data-centric routing. The MAC protocol must meet two goals. The first is to create a network infrastructure, which includes establishing communication links between may be thousands of nodes, and providing the network self-organizing capabilities. The second goal is to fairly and efficiently share communication resources between all the nodes. Existing MAC protocols fail to meet these two goals because power conservation is only a secondary concern in their development. Also, wireless sensor networks have no central controlling agent and a much larger number of nodes than traditional ad hoc networks. Any MAC protocol for wireless sensor networks must also take into account the ever-changing topology of the sensor network due to node failure and redistribution. Since sensor nodes are usually operated by batteries and left unattended after deployment, power saving is a critical issue in WSNs. Many research efforts in the recent years have focused on developing power saving schemes for wireless sensor networks. Designing power efficient MAC protocols is one of the techniques that prolong the lifetime of the network. In addition to energy efficiency, latency and throughput are also important features for consideration in MAC protocol design for WSNs. Commercial standards like IEEE 802.11 have a power anagement scheme for ad hoc networks, wherein the nodes remain in idle listening state at low traffic to conserve power, significant power is wasted even in the idle listening mode. Hence, IEEE 802.11 is not suitable for sensor networks. A properly designed MAC protocol allows the nodes to access the channel in a way that saves energy and support QoS.

Network Layer:

The network layer in a WSN must be designed with typical considerations in mind, ever existing power efficiency, WSNs are data-centric networks, and WSNs have attribute-based addressing and location awareness. The link layer handles how two nodes talk to each other, while the network layer is responsible for deciding which node to talk to. The simplest design is flooding. When using flooding, each node receiving data repeats it by broadcasting the data to every neighbor unless the max hop lifetime of the data has been reached or the receiving node is the destination. The major support for flooding is the simplicity. It requires no costly topology maintenance or complex route discovery. The shortcomings, however, are substantial:

- Implosion, it occurs when two nodes (A and B) share multiple (n) neighbors. Node A will broadcast data to all n of these neighbors. Node B will then receive a copy of the data from each of them.

- Overlap, when two nodes share the same sensing region. If a stimulus occurs within this overlap, both nodes will report it.

- The last and most crucial problem is resource blindness. Flooding does not take into account available energy resources. Gossiping is an enhancement to flooding. In gossiping, when a node receives data, it randomly chooses a neighbor and sends the data to it. Gossiping avoids the problem of implosion, but does not address the other two concerns and contributes to the latency of the network.

A step up from flooding and gossiping is ideal dissemination. In this algorithm, data is sent along a shortest-path route from the originating node. Such approach guarantees that every node will receive every piece of information exactly once. No energy is wasted in sending or receiving redundant data. However, the overhead involved in keeping track of the shortest paths is substantial. Also, ideal dissemination does not take into account that some node may not need a particular piece of information; nor does it allow for resource awareness.

A little more sophisticated family of protocols is sensor protocols for information via negotiation (SPIN). The SPIN family addresses the deficiencies of classic flooding by negotiation and resource adaptation. With more sophisticated and energy aware techniques for data dissemination, it reduces the amount of energy expended, solves the problems of implosion, overlap, and resource blindness, and ensures that only interested nodes will expend energy to receive data. Negotiation helps to overcome the problems of implosion and overlap and ensures only useful and desired information is disseminated.

In order for negotiation to work, nodes must describe the data to be sent using meta-data. In order for SPIN to be efficient the meta-data must be significantly shorter than the data being described. Also, meta-data describing two distinguishable pieces of data must be different. Likewise, if two pieces of data are indistinguishable, they will share the same meta-data. The format of the meta-data is not specified by SPIN, but rather application specific. SPIN-2 is an implementation of SPIN that employs a low-energy threshold. When energy is abundant, the node functions as normal. However, when the resource manager detects that a node power supply is reaching the low- energy threshold, the node will not participate in later stages of the protocol. This prolongs the life of the node and allows it to perform only high priority functions. SPIN is a more sophisticated and energy aware schema for data dissemination. It reduces the amount of energy expended, solves the problems implosion of, overlap, and resource blindness, and ensures that only interested nodes will expend energy to receive data.

Transport Layer:

Transport control protocol for WSNs should account for several concerns:

- Congestion control and reliability. The more data streams flow from sensor nodes to sinks in WSNs, the more congestion might occur around sinks. Also, there are some high-bandwidth data streams produced by multi-media sensors. Therefore, it is necessary to design effective congestion detection, congestion avoidance, and congestion control mechanisms for WSNs. Although MAC

Protocol can recover packets loss from bit-error, it has no way to handle packets loss from buffer overflow. Then the transport protocol for WSNs should have mechanism for packets loss recovery such as ACK and Selective ACK as used in TCP protocol so as to guarantee reliability.

Reliability under WSNs may have different meaning from traditional networks that generally guarantee correct transmission of every packet. For some application, WSNs only need to correctly receive packets from a certain area, not from every sensor nodes in this area, or may be contempt with some ratio of successful transmission from a sensor node. These modified reliability concept motivates the design of different transport control protocols. It would be better to use hop-by-hop mechanism for congestion control and loss recovery since it can reduce packet dropping and conserve energy.

The hop-by-hop mechanism can simultaneously lower buffer requirement at intermediate nodes, which suits the limited memory sensor nodes.

- Simplifying initial connecting process or use connectionless protocol so as to speedup start and guarantee throughput and lower transmission delay. Most of applications in WSNs are reactive, that is passively monitor and wait for event occurring before reporting to sink. These applications may have only few packets for each reporting, and the simple and short initial setup process is more effective and efficient.

- Avoiding packets dropping as possible to lessen energy wastage. In order to avoid packet dropping, the transport protocol can use active congestion control at the cost of a lower link utility. The active congestion control (ACC) can trigger congestion avoidance before congestion occurs. An example of ACC is to make sender (or intermediate nodes) reduce sending (or forwarding) rate when the buffer size of their downstream neighbors overruns a threshold.

- Guaranteeing fairness for different sensor nodes so that each sensor node can achieve a fair throughput. Otherwise, the loaded sensor nodes cannot properly report events in their area, which leads to erroneous monitoring, tracking, and control.

- Enabling cross-layer interaction. If a routing algorithm can notify route failure to the transport protocol, the transport protocol will know that packet loss is not from congestion but from route failure, and consequently the sender will regulate its current sending rate to guarantee high throughput and low delay.

Application Layer:

To address application layer protocols, it is primordial to address some functions that are to be implemented, specifically, data fusion and management, clock synchronization, and positioning. A WSN is intended to be deployed in environments where sensors can be exposed to circumstances that might interfere with provided measurements. Such circumstances include strong variations of pressure, temperature, radiation, and electromagnetic noise. Thus, measurements may be imprecise in such scenarios. Data fusion is used to overcome sensor failures, technological limitations, and spatial and temporal coverage problems. Data fusion is generally defined as the use of techniques that combine data from multiple sources and gather this information in order to achieve inferences, which will be more efficient and potentially more accurate than if they were achieved by means of a single source. The term efficient, in this case, can mean more reliable delivery of accurate information, more complete, and more dependable. The data fusion can be implemented in both centralized and distributed systems. In a centralized system, all raw sensor data would be sent to one node, and the data fusion would all occur at the same location.

System administrators interact with WSNs sensor management protocol (SMP). Unlike many other networks, WSNs consist of nodes that do not have global IDs, and they are usually infra- structureless. Therefore, SMP needs to access the nodes by using attribute-based naming and location-based addressing. SMP is a management protocol that provides the software operation needed to perform several administrative tasks.

- Introducing to the sensor nodes the rules related to data aggregation, attribute-based naming, and clustering.

- Exchanging data related to the location finding algorithms.

- Time synchronization of the sensor nodes.

- Moving sensor nodes.

- Turning sensor nodes on and off.

- Querying the sensor network configuration and the status of nodes, and re-configuring the sensor network.

- Authentication, key distribution and security in data communications.

3.6 Cross-Layer Protocols for WSNs

The severe energy constraints of battery-powered sensor nodes necessitate energy-efficient communication protocols in order to fulfill the application objectives of WSNs. It is much more resource-efficient, according to some research, to have a unified scheme which melts common protocol layer functionalities into a cross-layer module for resource-constrained sensor nodes. A unified cross-layer communication protocol, for efficient and reliable event communication, considers the effects on WSNs of replacing transport, routing, medium access functionalities, and physical layers (wireless channel). A unified cross-layering is such that both the information and the functionalities of traditional communication layers are melted in a single protocol. The objective of the proposed cross-layer protocol is highly reliable communication with minimal energy consumption, adaptive communication decisions and local congestion avoidance. Protocol operation is governed by the concept of initiative determination. Based on this concept, the cross-layer protocol performs received based contention, local congestion control, and distributed duty cycle operation in order to realize efficient and reliable communication in WSN. Performance evaluation reveals that the proposed cross-layer protocol significantly improves the communication efficiency and outperforms the traditional layered protocol.

•●•

UNIT - 4

Localization And Routing

In WSN nodes, localization provides important piece of information-the current location of the node. This chapter concentrates on localization and explores what types of location information exist for WSN and how to obtain them with some basic techniques. If you are already proficient in localization techniques for other applications, do not skip this chapter, as localization in WSN can be very different.

4.1. Localization Challenges

Location is for humans something as intuitive and natural as time. Just recall one of the typical human nightmares of not knowing where you are and what time we have. For devices, knowing the location is not that intuitive and easily achievable. Most of devices do not know where they are, ranging from washing machines and coffee makers to cars and laptops. Only when a GPS receiver is installed, does a device have location information to use. The following sections discuss the properties and peculiarities of location information for sensor nodes.

4.1.1 Types of Location Information

First, let us explore what kind of location information is needed. Global addresses, such as postal addresses, are typical for outdoor environments. When these are not available (outside of the postal system), then GPS coordinates are required. Such location information is also very useful for outdoor sensor networks. But what happens if the complete sensor network is in the same building or building complex? GPS is not available indoors and the postal address is too coarse because it will be the same for all nodes. In this case, you must use semantic information such as the floor or room number of the sensor node. Where should we look for the sensor? Perhaps in the entrance or somewhere completely else?

Other locations may be too precise for some applications, such as the location "desk" for the sensor in building ZHG. There also might be sensors located on different professors' desks but how can you differentiate between them? Furthermore, you need to note the difference between *symbolic and physical location*. Physical coordinates remain legal at all times, e.g., GPS coordinates. Symbolic coordinates, such as "desk" can easily change over time, either because the desk is moved away together with the sensor or the desk is moved away, but the sensor node remains where the desk was and gets a new symbolic position, e.g., "chair". Postal addresses are also considered symbolic, even if street names, for example, do not change frequently.

However, the largest problem with postal addresses is their language dependency. For example, in English the country is called Germany, but in German it is called

Deutschland. This problem is also defined as the *scope* of the *location*. If the scope is a single room, then "desk" is perfectly acceptable. If the scope is the campus, then the room number plus the prefix of the building (e.g., room MZH 1100) works fine. If the scope is the world, you would either need to add the address of the building or the GPS coordinates. It is important to decide what the scope is and the kind of location information you need before you start implementing the localization method.

4.1.2 Precision against Accuracy

Different from location scope, localization precision and accuracy refer to how well you are able to localize the sensor nodes. The definition of localization accuracy is the largest distance between estimated and real position of the sensor node.

As an example, if you use room numbers as location information, you can have accuracy of one room, several rooms, or even several floors. Or, if you use GPS coordinates from GPS receivers, the accuracy is typically 15 meters. This means that any location estimation is expected to be in a radius of 15 meters around the real position.

4.1.3 Costs

Different localization methods have various costs. Some have purely *financial costs*, such as installing a GPS receiver at each individual sensor node. Others have *space costs*, for example, when the GPS receiver cannot fit into the planned space for the sensor node. These examples are often found in industrial settings, in which nodes have to fit in very small elements such as tubes or machines. Other possible costs include *communication* or *energy costs*, where the method requires so much communication between the sensor nodes that it becomes unbearable, especially when nodes move around. *Infrastructural* costs refer to the special installation of some infrastructure to localize the nodes, such as GPS anchors.

4.2 Pre-Deployment Schemes

Pre-deployment schemes include all possible methods to provide sensor nodes with their location information before installing them in the environment. This is very often done for symbolic information, such as furniture items, room labels, or postal addresses. The process is simple yet very error-prone. Every node gets its location manually and you cannot change it later. However, very good deployment planning is needed, where exactly the planned nodes are positioned exactly at the planned positions. Obviously, this method does not scale very well, as it becomes very time consuming to program the locations of hundreds of nodes. Installing individual GPS receivers on each sensor node is also a kind of predeployment solution. In this case, the accuracy is very good (at least for outdoor and relatively large deployments) but the financial cost for acquiring the devices is high. Furthermore, the energy cost for running these devices is also high and often not affordable.

4.3 Proximity Schemes

Let us now turn to post-deployment localization. A simple technique is to guess the approximate location of a node, provided its neighborhood. This approach. There is an installed *anchor* in each room, which provides everybody who can hear it with its location. Thus, a sensor node can explore its neighborhood for anchors and if it sees one, it can assume that it is close enough to have the same approximate location. For example, node 1 sees the anchor of room A and can assume that it is also in room A. The method is not a very accurate. It can also incur various problems and conflict situations. For example, node 2 sees only the anchor of the corridor and assumes it is there instead of the correct location in room D. Node 4 sees three different anchors and will have trouble deciding for one of them. Node 3 does not see any anchor and cannot position itself at all. Various remedies have been implemented to tackle the previous problems.

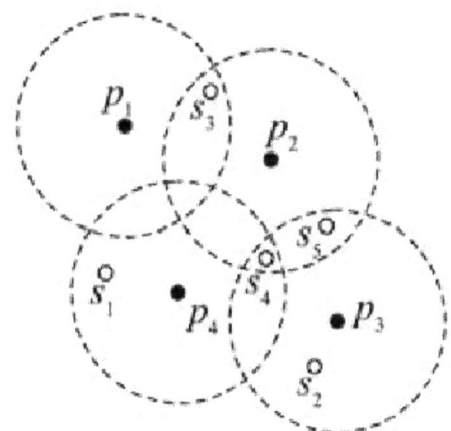

Figure 4.1. Proximity localization for sensor nodes

4.4 Ranging Schemes

Ranging essentially means measuring the location. For sensor nodes localization, you first do some measurements and then decide on your location. Again, we use anchors and their locations. However, instead of adopting their location as in proximity schemes, in ranging schemes we try to compute our own position depending on the measurements that we do.

Triangulation and *trilateration* are the two main approaches (Figure 4.2). In triangulation, you measure the angle between the sensor node and the anchors and use this information to compute your own position. In trilateration, you measure the distances to the anchor nodes to compute your own position. The mathematical details for both computations are quite complex so they are not covered here but the interested reader can read Karl and Willig's book [1]. Here, discussion focuses on both approaches' important properties and consequences.

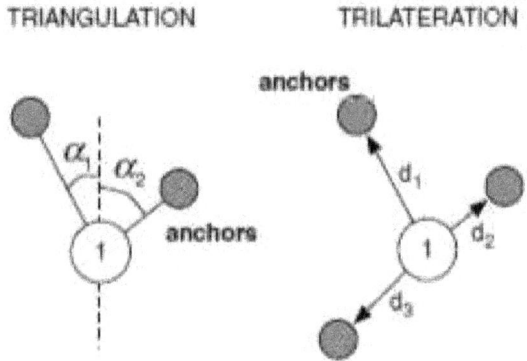

Figure 4.3 Two Ranging schemes

Triangulation:

First, you need to answer the question of how to obtain angle measurements. This is not that simple with standard sensor node hardware, which typically uses omnidirectional radio transceivers. Special hardware is required, such as an array of antennas or microphones on different sides of the sensor node to understand from which *direction* the signal arrives. This approach is costly in terms of hardware but can achieve quite good accuracy. In terms of the calculation needed, it is relatively simple and does not require any special or costly mathematical functions. It needs a minimum of two anchors for a two-dimensional space.

Trilateration:

You have already read about some attempts to measure distance based on radio communication such as mapping RSSI values into distances and know this approach is not very reliable. However, others do exist. For example, one could use two different communication interfaces on-board such as radio and acoustic. If the anchor sends a pulse simultaneously through each of the interfaces, neighboring sensor nodes receive them slightly separated in time. Given the propagation speed of both pulses (speed of light for the radio and speed of sound for the acoustic interface), the sensor node can accurately calculate the distance to the anchor. This method is called *time difference of arrival* (Figure 4.3). Keep in mind, the larger the time difference of arrival, the greater the distance to the anchor. Again, this may require additional hardware, which is costly, but not as costly as an array of antennas. In terms of computation, you need more resources to compute the location of the sensor nodes and a minimum of three anchors.

4.4.1 Range-Based Localization

How does a localization protocol ranging has sensor nodes that look procedure. If they gather enough anchor points and measurements, they compute their own locations. Once they have localized, they also declare themselves anchors. In this way, sensor nodes without enough anchors get a chance to localize. This approach is also called *iterative localization*. The accuracy of localization, **localization** decreases significantly with each iteration when new anchors are added because the error sum ups over iterations. However, in this way, location information is available at a low price, high scalability, low communication and processing costs.

4.4.2 Range-Free Localization

Range-free localization is a combination of proximity - based and ranging techniques. Strictly speaking, instead of measuring exact distances or angles, this method means trying to guess approximate values and use those for calculating the sensor node's location. The following sections explore some of the most important and widely used strategies.

4.4.2.1 Hop-Based Localization

In Figure 4.4, three anchors are present in the network and you have their exact positions. Instead of using real distances to the anchors, you can try to approximate how long one hop in this network is. What you already know is the number of hops between the anchors (the bold links in the network) and the real distance between them. For example, the distance between anchors A and B is 130 meters, and they are 5 hops separated. This results in a mean length of 1 hop of 26 meters. The distance between anchors A and C is 170 meters and 6 hops, resulting in a mean hop length of 28.3 meters. The last available distance is between anchors B and C, which is 120 meters and 5 hops, resulting in a mean hop length of 24 meters. You can see that the results are quite close to each other and the network-wide mean hop length can be calculated as (24+28.3+26)/3 = 26.1 meters. The localization for the sensor nodes works then as follows. It calculates how many hops away it is from each anchor. It uses the mean hop length to calculate the approximate distance to this anchor and can perform trilateration to obtain its own location. Looking at the topology and real distances in Figure 4.4, it is clear that this approach is not very accurate. The savings from ranging techniques are also not significant, as simple techniques such as RSSI measurements to neighbors are less costly than the mean hop length calculation. However, for really large and dense networks, this approach is well suited.

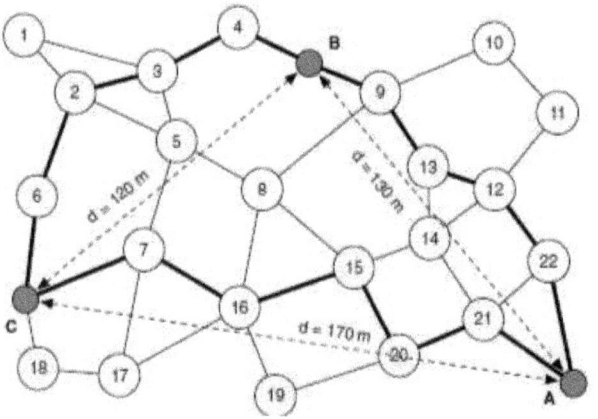

Figure 4.4 Hop bases Sensor localization structure

4.5. Routing Basics

There are several scenarios for multi-hop communication or routing. They are differentiated based on the source and the destination of the data inside the network. First, we will explicitly define source and destination, as well as some important special roles of the nodes in the network.

- **Data Source:** It is a node, which produces the required data and is able to send it out to other nodes in the network.

- **Data Destination: It** is a node, which require the data and is able to receive it from other nodes in the network.

- **Data Forwarder:** It is any node in the network, which is not source or destination of the data, which is able to receive the data from another node and send it to further in the network.

- **Data Sink:** A dedicated node (s) in the network, which is the destination of any data in this network.

- **Full-network broadcast.** There is a single source of the data and all nodes in the network are implicit destinations of the data.

- **Unicast.** There is a single source and a single destination of the data, which could be any nodes in the network.

- **Multicast.** There is a single source, but multiple destinations of the data. Again, both the source and the destinations can be any nodes in the network.

- **Converge cast.** All nodes in the network are sources, while a single node (typically the sink) is the destination.

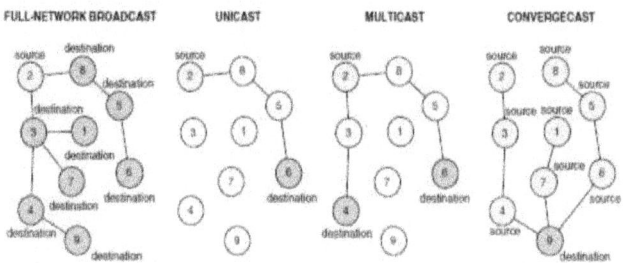

4.5.1 Routing Metrics

The routing metric is the driving force behind routing. It represents a way to compare different sensor nodes in terms of their vicinity to the destination. This vicinity can be a precise or approximated geographic vicinity, but it can also be a communication-based vicinity such as least delay.

4.5.1.1 Location and Geographic Vicinity

Geographic location is probably the most logical metric to compare two nodes in terms of their vicinity to the destination. Real geographic locations can be used, but also approximations such as distances or similar. The main requirement is that all nodes are aware of their own position and that they have the position of the destination (recall that this is communicated by the destination itself via a full network broadcast) as shown in figure 4.5.

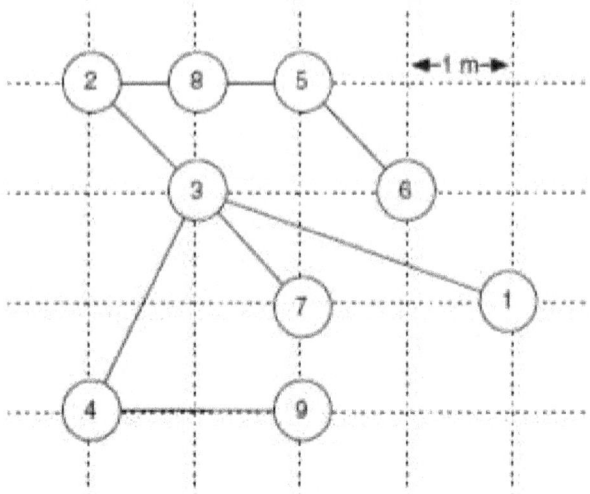

Figure 4.5 Sample Network

4.5.1.2. Hops

Another way of approximating location and vicinity to the destination is to use number of hops. Here, the assumption is that a hop corresponds to some distance in the network, thus more hops correspond to longer distance... This idea is visualized in Figure 4.6. Nodes that can directly reach the destination are one hop away from it; nodes that can directly reach any one-hop node are two hops away from the destination. The number of hops is correlated with the real geographic distance but also takes into account the network topology and the available links in it. For example, node 5 in Figure 4.6 is closer geographically to the destination than node 9, but one hop further away than node 9. This is important, as geographic proximity does not guarantee the availability or the shortness of a route to the destination. Hop-based routing exhibits some further advantages and disadvantages compared to geographic routing. First, the number of hops from the destination is not readily available at all nodes, but it needs to be discovered first. For this, usually the destination sends a full network broadcast to all nodes.

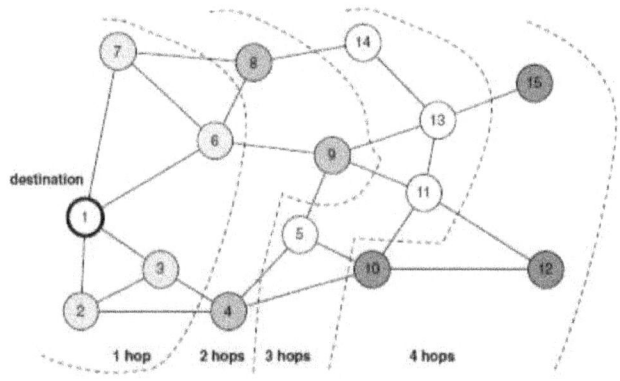

Figure 4.6 Hops Mechanism

4.5.1.3 Number of Retransmissions

In the best-case scenario, the number of hops also represent how many times a packet needs to be forwarded to reach the destination. However, in reality, a packet often needs to be retransmitted several times before it reaches even the next hop. The main reason is interference, but also bad links over long distances or too many obstacles can cause packet loss and retransmission. To better represent this behavior in the routing metric, you can use the real number of retransmissions over the route, instead of the hops.

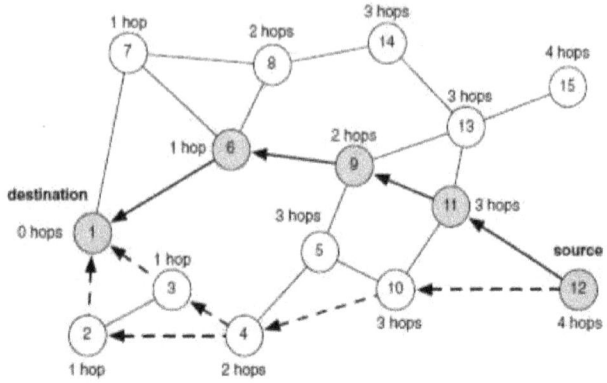

Figure 4.7 Retransmission mechanis

4.6 Routing Protocols

A routing protocol is an algorithm, which defines how exactly to route the packet from the source to the destination. It uses one or even more of the previous metrics to evaluate the network conditions and to decide what to do. The following explores basic but important protocols for routing in wireless sensor networks. Even if they seem different from each other, they all attempt to comply with the main routing protocol requirements for WSNs:

Energy efficient. Protocols need to be able to cope with node sleep and to have little overhead for route discovery and management.

Flexible. Protocols must be able to cope with nodes entering or exiting the network (e.g., dead nodes or new nodes) and with changing link conditions.

4.6.1 Full-Network Broadcasts

The simplest way to send a packet to any node in any network is to send the packet to all nodes in the network. This procedure is called flooding or full-network broadcast.

4.6.2 Location-Based Routing

In this approach, the routing metric is **location**, i.e., the geographic coordinates of the nodes. The data source is aware of the geographic coordinates of the data destination. Additionally, any node of the network is aware of its own geographic coordinates and is able to ask its neighbors for their coordinates. It is not necessary to know all coordinates of all nodes or to have real geographic coordinates such as longitude and latitude. Instead, relative coordinates on a local system are sufficient, as far as all nodes are using the same system.

4.6.3 Directed Diffusion

Directed diffusion is the name of a routing protocol, proposed by Intanangonwiwat *et al.* at the beginning of the sensor networking era around 1999 to 2000. It is a very particular routing mechanism because it does not define information sources or destinations. Instead, it uses a paradigm called *information interest* and a publish subscribe mechanism to deliver data to many different interested destinations.

•●•

UNIT - 5

Data Aggregation and Security

5.1 Clustering Techniques

Clustering was one of the first techniques applied to very large networks. The main idea is to organize the network into smaller sub-networks, so that data can be collected and analyzed in a location-restricted way, meaning only some important or aggregated data comes to the final network-wide sink.

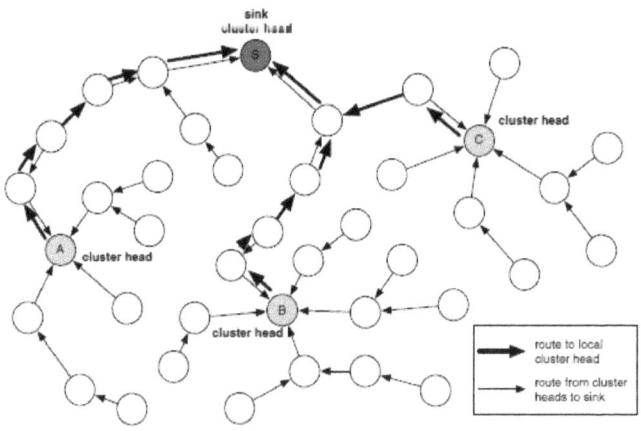

Figure 5.1 Cluster Network Example

Figure 5.1. In clustering, we differentiate between the following roles of individual nodes:

- Cluster members are simple nodes, which sense some phenomena. They send their data to their cluster heads. A cluster member belongs to a single cluster head.

- Cluster heads are local sinks, which gather all of the information of their cluster members. Several options are possible here and depend on the application: send

everything through a more powerful link to a final data storage place; aggregate or compress the data then send it through the sensor network to a global sink; or analyze the data and either report some event or not.

- The sink, sometimes called the global sink, is a single node in the network, which gathers all information from all nodes in the network. It is not necessary that it exists at all. What is the difference between clustering a large network and deploying several small ones? When deploying several small networks, they will be independent of each other and at the same time the engineer needs to make sure that they do not interfere with each other. This approach is rather cumbersome and inflexible. In a large network, you can take advantage of all nodes and allow the nodes to cooperate to achieve their goal.

The challenge is that the network is too big to be handled by normal data collection protocols. The main objective of clustering is to save energy and to avoid data congestion in the network. The underlying assumption is that if you try to route all of the data of all nodes to a single sink, then at least some of the nodes will be completely congested (the ones close to the sink) and that their batteries will be drained too fast. Furthermore, it is important which are the cluster heads and how they are selected and managed.

5.2 In-Network Processing and Data Aggregation

There are mainly two concepts for in-network aggregation and processing: compression and aggregation. In compression, data remains as it is, but its resource usage is minimized. For example, 5 packets with 5 bytes data in each could be easily combined in a single packet and this would save the communication overhead of sending 5 different packets. In aggregation, the data is somehow processed and only part of it continues to the sink. For example, only the maximum sensed value is allowed to travel to the sink and all lower values get discarded on their way. The following sections discuss some examples.

5.2.1 Compression

The first choice for compression is combining data from several packets into a single one. This concept can always be used and delivers very good results. However, it also relies more or less on the arrival of packets at the sink and thus cannot be used for real-time applications. Let us explore the example in Figure 5.2. On the left, you see a sample network, where data is routed with the help of the Cluster Tree Protocol routing protocol. Without compression, nodes 4, 7, and 10 would need to forward 2 packets every sampling period-their own packet and one from a child node. Node 5 would need to forward 3 packets and node 3 even 4 packets. This sums up to a total of 17 transmissions.

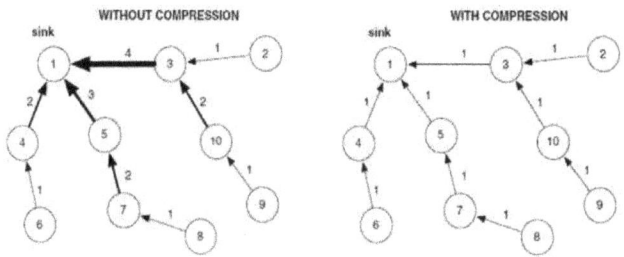

Figure 5.2 Comparison between without Compression and with Compression

If we simply combine the packets at each hop into a single one, each node would have to transfer only 1 packet, which sums up to only 9 transmissions, exactly as many as nodes exist in the network. Recall that transmissions are costly because each of them needs time to obtain the wireless medium and require not only the data to be transferred, but also various protocol headers, control packets, acknowledgements, etc.

Huffman Codes There is also another potential problem. Sometimes the data is so large that several pieces of it do not fit into a single packet. Then, you need to introduce some real compression techniques to minimize its size and to fit more of it into a single packet. Here, data compression algorithms come into play such as Huffman codes. A Huffman code assigns a special code to each character, which is shorter than its original encoding. Let us explore an example. Assume that you have an alphabet with 5 characters only.

To encode those with bits, you need at least 3 bits. The codes for these characters could be 000, 001, 010,011, and 100. The Huffman code calculates first the probability of how often individual Characters occur. For example, after analyzing some sample text in this alphabet, you can identify the probabilities for your characters as 0.1, 0.3, 0.2, 0.15, and 0.25. This is depicted in Figure 5.2. Now, you can create the Huffman tree. For this, the Characters are ordered by their probability and then the two lowest ones are connected with a next level node, until the root is reached. Now, you can start assigning codes to the individual nodes. Assign a 0 to the node with the character b (the one with the highest probability) and to everything else a 1. Next, assign a 0 and 1 to the next level of nodes and you get 10 for the character e and 11 for everything else, and so on.

5.3 Security issues in Wireless Sensor Networks

Network security is one of the most pressing concerns in all wireless networks, including wireless sensor networks. In this section, we briefly introduce the security problem and explain some of the specifics of wireless sensor networks.

Fundamentals:

Network designers have to be aware of and decide about suitable mechanisms to implement one or more of the following general security goals

- **Confidentiality** Information should only be revealed to authorized entities; any other entity should not be able to discover the information from eavesdropping or from reading memories.

- **Data integrity** The receiver of information wants to be sure that it is not modified in transit, either intentionally or by accident. To distinguish unmodified "wanted" information from unmodified bogus information, the originator must be identifiable uniquely.

- **Accountability** The entity requesting a service, triggering an action, or sending a packet must be uniquely identifiable.

- **Availability** Legitimate entities should be able to access a certain service/information and to enjoy proper operation.

5.4 Denial-of-service attacks

consider a number of different denial-of-service attacks in sensor networks, working at different levels. Denial-of-service attacks in general can try to (i) disable services, or (ii) to deplete service providers, for example, by overusing the service. To disable a sensor network's service, an attacker might simply destroy nodes. Although sensor networks have some resilience to node failures, the attacker can distort the network by destroying a large number of nodes or by focusing on especially important nodes, for example, sensor nodes in the vicinity of sinks that are needed for forwarding. In the following, however, we discuss protocol-related attacks in different layers.

- **Physical-layer and link-layer attacks:**

With physical-layer jamming, an attacker simply distorts radio communication. One way to achieve this is to place attacker nodes somewhere into the network and let them continuously send radio signals in the sensor network's frequency band. Especially effective is such an attack when the attacker nodes are close to sink nodes, effectively reducing a user's ability to control the network or to acquire data from it. A single attacker node can distort many neighbors at once and, by strategically placement of a number of attacker nodes, the whole sensor network can be disabled. A cleverer attacker can take knowledge about the protocols into account to save energy, giving rise to link-layer jamming. Especially, the MAC protocol is a good candidate. Let us consider, for example, protocols based on exchange of RTS/CTS Whenever an attacker node a receives an RTS packet issued by some node x, it can answer with a jamming signal, interfering with any CTS packet sent to x. As a consequence, x has no transmit opportunity, backs off and tries again later with another RTS packet. The attacker might exploit the MAC protocol further to save energy.

- **Network-layer attacks:**

Several types of attacks can be executed on the network layer. First, attacker nodes can behave similar to normal nodes; specifically, they can participate in routing protocols or dissemination of interests with the goal of directing routes to itself and to drop packets later on. This attack is called **black hole** attack. For example, in distance-vector protocols, the attacker can pretend to have particularly good routes to the sink. Dropping of packets destroys information, and furthermore, the forged route advertisements attract lots of traffic around the attacker, causing increased congestion levels and contention. In a similar kind of attack, so-called **misdirection's**, the adversary creates wrong routes, for example, by sending wrong route advertisement packets or by falsely answering route request packets. A wrong route can, for example, contain a loop and cause waste of energy. Another possible effect is that traffic does not reach the intended sink nodes. Instead of creating wrong routes, an adversary can also cause creation of unnecessary routes, for example, by issuing route lookup requests.

- **Transport layer and application attacks:**

If the transport layer uses explicit connections between identifiable nodes, either end of the connection needs to maintain some form of connection control block (CCB). Similar to TCP syn flood attacks, an attacker can issue a large number of connection setup requests and cause exhaustion of memory at the end nodes because of large numbers of unneeded CCBs.

5.5 Sensor Network Security properties

- *Semantic security*: Since the counter value is incremented after each message, the same message is encrypted differently each time. The counter value is long enough never to repeat within the lifetime of the node.

- **Data authentication**: If the MAC verifies correctly, a receiver can be assured that the message originated from the claimed sender.

- **Replay protection**: The counter value in the MAC prevents replaying of old messages. Note that if the counter were not present in the MAC, an adversary could easily replay messages.

- **Weak freshness**: If the message verified correctly, a receiver knows that the message must have been sent after the previous message it received correctly (that had a lower counter value). This enforces a message ordering and yields weak freshness.

- **Low communication overhead:** The counter state is kept at each end point and does not need to be sent in each message.

•●•

UNIT - 6

Designing And Deploying WSN Applications

6.1 Early WSN Deployments

Many researchers have already identified and discussed the problem of deployment. Some surveys also hold more information about recent successful deployments. There are two prominent examples for failing deployments: the so-called Potatoes application and the Great Duck Island application. Both applications are from the beginning of the sensor networks era, but they are useful to analyze caveats to inexperienced practitioners.

6.1.1 Murphy Loves Potatoes

This application, officially called LOFAR-agro, was developed for precision agriculture for potato fields in the Netherlands to monitor temperature and humidity, and to take centralized irrigation decisions based on the acquired data. The resulting publication is worth reading, especially because the authors describe in an honest and direct way their disastrous experience from a project, which at the end managed to get only 2% of the targeted data for a much shorter period of time than targeted. The problems were numerous. After careful testing of the nodes in the lab, an accidental last-minute update to the code of the MAC protocol caused it to only partially work and not deliver any data at all. Yet another problem was caused by the code distribution system installed, which caused the nodes to continuously distribute new code throughout the network and depleted the nodes' batteries in only 4 days. The next revision brought up a problem with routing, as it was trying to maintain a table of all neighbors, sometimes up to 30. However, such large routing tables were not planned and could not be stored, resulting sometimes in no routing paths available to the sink. An attempt to manage node reboots wisely and the addition of a (faulty) watchdog timer resulted in most of the nodes rebooting every couple of hours. In summary, instead of running for more than 6 months (the complete potato growing period), the network ran for only 3 weeks and delivered 2% of the gathered data during this short period. Node reboots, lifetime problems, and faulty code were the problems. in this example.

6.1.2 Great Duck Island

This is one of the first-ever WSN applications for environmental monitoring. In 2002, 43 nodes were deployed on Great Duck Island, USA. The nodes were equipped with infrared, light, temperature, humidity, and pressure sensors. Once installed, researchers were not allowed to go back to the island because they would disturb the island birds while they were breeding. Again, the problems were numerous, but this time caused mainly by weather and environmental conditions resulting in hardware faults. A crash of the gateway caused the complete data of two weeks to be lost forever. Water entering the casing of the nodes interestingly did not cause them to fail, but for the sensors to deliver faulty data such as humidity readings of over 150%. These errors triggered an avalanche of problems with the other sensors because they were all read through the same analog digital converter.

However, communication problems were actually rare because of the network being a one-hop. In 2003, a second attempt was made but routing was introduced and the casings were updated. The trial was also divided into two batches with 49 nodes deployed as a one-hop network, similar to the year before. Batch B included also routing and consisted of 98 nodes. Again, the gateway failed several times causing much of the data to be lost. The node hardware performed better this time, but routing caused overhearing problems at the nodes, which was not previously tested in the lab. This caused many nodes to die prematurely. In the end, the first 2002 deployment Delivered approximately 16% of the expected data and the second trial delivered 70% for the batch A network but only 28% for batch B. The ratio of gathered data against Expected data is also called the *data yield*. **Data yield** Note here the different set of problems.

Hardware issues and maintenance unavailability were great enemies. Furthermore, the researchers added unnecessary complexity to the network such as the unneeded routing protocol. Semantic problems are one of the most difficult problems for WSN practitioners, as the data gets delivered successfully but it is completely useless. This typically renders all of the data useless, as no way exists to identify which readings are faulty and which are not, so they all need to be considered faulty.

6.2 General Problems

Looking at the experiences of the WSN researchers from the previous examples, what is the definition of a problem? A problem is an unexpected behavior of a node or a group of nodes, which does not correspond to your formal or informal specification.

The following differentiates problems in more detail,

- **Node problems** occur on individual nodes and affect their behavior only.
- **Link problems** affect the communication between neighbor nodes.
- **Path problems** affect the whole path of communication over multiple hops.
- **Global problems** affect the network as a whole.
- These problems are caused by varied processes and properties in the environment, but mainly by the following:
- **Interference** in the environment causes various communication problems in the network as a whole.
- **Changing** properties of the environment, such as weather conditions, furniture, or people density, causes various communication problems but also other hardware problems such as complete shutdowns due to low temperature.

- **Battery** issues cause not only problems at the node level but also for communication and sensing as hardware performance deteriorates slowly with decreasing battery levels

- **Hardware reliability** causes various global problems such as communication issues, sensing issues, but also node reboots.

- **Hardware calibration** is a typical problem for sensing because it delivers wrong sensor data.

- **Low visibility** of the nodes' internal states does not directly cause problems but Favors unrecognized errors until late in the design process, which prove to be hard to detect and resolve. The following sections explore the individual problem levels in more detail, before we discuss how to avoid them in the design and deployment process.

6.2.1 Node Problems

Node death *Node death* is one of the most common problems. Suddenly, a node stops working completely. This might seem like a serious problem and it is, but there is also some good news. When a sensor node suddenly stops working, its internal state is still valid. This means that whatever data already exists on this node, it is still there and valid (unless the sensor does not store data at all). Also, neighbor nodes are usually well prepared for dead nodes and quickly exclude them from routing or co-processing. Thus, real node death is a problem, but a manageable one. However, what sometimes looks like a dead node is often not a completely dead one. This can be due to various reasons such as:

- **Slowly deteriorating batteries** causing individual components on the node to perform badly or not at all, while others continue working.

- **Faulty components** causing errors while sensing, storing, or communicating, while the general behavior of the node is not affected. Faulty components could be both hardware and software.

- **Node reboots** typically take time and cannot be easily detected by neighbors. They are often caused by software bugs. During this time, the node seems to be unresponsive and uncooperative.

These problems lead to unpredictable node behavior, which results in semantic errors in the application. This means that the data coming out of the network does not make sense any more. These errors are extremely hard to detect and localize, as they **seemingly** are unpredictable and in deterministic. Thus, the *seemingly dead nodes* are by far the **dead nodes** largest problem in WSN deployments. More sophisticated strategies can also track the state of individual hardware and/or software components such as radio or sensors. They are often programmed to report problems by returning some sort of an error (e.g., false to a send packet request of the radio). However, these more sophisticated techniques require more memory and processing and they can also overwhelm the primary application. Other node problems include various software bugs. These lead to clock skews, which lead to unsynchronized nodes and missed communication opportunities, hanging or killed threads, overflows in counters, and many other issues. It is essential to test the software incrementally and thoroughly to avoid the manifestation of these problems at the deployment site.

6.2.2 Link/Path Problems

Obviously, node problems also cause link and path problems. General propagation or path loss is also a problem because nodes are prohibited to communicate where expected. However, the biggest issue is *interference*. Interference can occur between **interference** individual nodes in your network but also between your nodes and external devices such as mobile phones, Bluetooth devices, or a high-voltage power grid. You cannot see the interference and it changes quickly and unpredictably. In fact, interference by itself causes "only" lost packets. However, these sporadically lost packets trigger an avalanche of other problems at all levels of the communication stack. They cause more traffic because packets need to be resent. They also cause link prediction fluctuations because the link protocols believe something happened in the network. They force the routing protocol to change its paths and cause the application to buffer too many packets, leading to dropped packets. At the routing level, these fluctuations might lead to temporarily unreachable nodes or loops in the routing paths. These, in turn, cause even more traffic and can make the network collapse completely.

How can you achieve robust communications? The employed protocols must be simple and leverage all the communication opportunities that they have. For example, a highly agile link quality protocol, which supports even high mobile nodes, may be a very interesting academic topic but will not work in reality. It will be too fragile for highly fluctuating links under heavy interference. At the same time, another link predictor, which is rather slow in recognizing new or dead nodes, will perform more robustly. At the same time, we should not forget why we always target better protocols and algorithms. They are able to save a lot of energy and significantly increase the network lifetime. As always, you need a tradeoff between robustness and energy saving. Other link problems also exist, including traffic bursts, which cause links to congest and packets to get dropped. Asymmetric links lead to fake neighbors. Both can be mitigated in the same way as general link and path problems by designing robust communication protocols.

6.2.3 Global Problems

These problems affect the network's general work and can be coarsely classified into topology, lifetime, and semantic problems as shown in Figure 6.1. Topology problems refer to various peculiarities in the communication topology of the network, which lead to errors or other unexpected behavior.

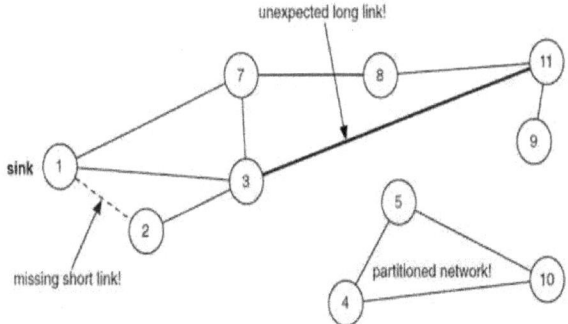

Figure 6.1 Topology related Global Problems, missing short links, unexpected long links and portioning of the networks

Missing short links are links between very close neighbors that are not able to communicate short links indicate with each other. These links are expected and planned by the designer to unexpected offer alternative communication paths. The opposite problem, unexpected long links, long links occur when two neighbors with an unexpected long distance between them are able to communicate well. Both problems cause premature battery depletion of some nodes. Another difficult topology problem caused by missing links is partitioned net partitioned works. In these networks, nodes in one partition are completely isolated from the network's nodes in other partitions. This causes the partition to spend a lot of energy in search of a path to the sink and never send its data.

These problems are depicted in Figure 6.1. The missing link between nodes 1 and 2 causes all the traffic of node 2 to go through node 3, instead of going directly to the sink. Thus, node 3 experiences congestion and will use its battery much faster than desired. The unexpected long link between nodes 11 and 3 causes even more traffic to pass through node 3. All of the traffic from node 9 and 11 will use node 3 instead of nodes 8 and 7. Finally, the partition of nodes 4, 5, and 10 cause these nodes to never deliver any data and waste their batteries searching for routes. Lifetime problems refer to an unexpectedly short network lifetime.

There are server network real definitions for network lifetime, but the one that makes most sense in the context lifetime of real applications is the time when useful data is gathered. This time might start Immediately after deployment or sometime after it. It ends when the network is not able to retrieve as much data as it is needs to semantically process it. Obviously, this definition will have different interpretations in different applications. For example, if you have a plant watering application, in which the humidity level of various pots is monitored and reported, the information remains useful even if some of the nodes do not work anymore.

However, if you have structural monitoring of a bridge, you need all sensors to be working properly and reporting data, otherwise the structural integrity of the bridge cannot be guaranteed. Lifetime problems occur most often when individual nodes use their energy too quickly and deteriorate the network's general performance. The most endangered nodes are the ones close to the sink or at communication knots (they serve as forwarders for many other nodes). However, software bugs also can cause them to stay awake for extended periods of time and waste their energy. Lifetime problems are hard to predict or test because they manifest themselves too late in the process.

6.3 General Testing and Validation

Whatever the purpose of an application or the target of a study, there are several testing Methodologies applicable to WSNs. The main principle of testing and validation is to:

Figure 6.2 depicts the typical life cycle of an application. The implementation process itself is divided into the following five steps:

1. Start on paper with theoretical models to validate the general idea and algorithms.

2. Proceed with the implementation and simulate the system.

3. Increase the complexity by moving from simulation to a test bed deployment.

4. Further stress the system by introducing all real-world properties in a test deployment.

5. Finally, you should be ready to face the normal operation of the system in its final deployment site.

The previous steps can be taken one-by-one or by skipping some of them. However, skipping any of them either makes the next step more complex or does not sufficiently stress-test the system and thus does not prove that it is ready for operation. Next, we will discuss of the previous life cycle steps in detail. To render the discussion more concrete and realistic, a simple application will be used as a sample scenario called vineyard monitoring. The targeted deployment is depicted in Figure 6.3. The sink is a bit further away in a small house. Each of the sensors in the vineyard.

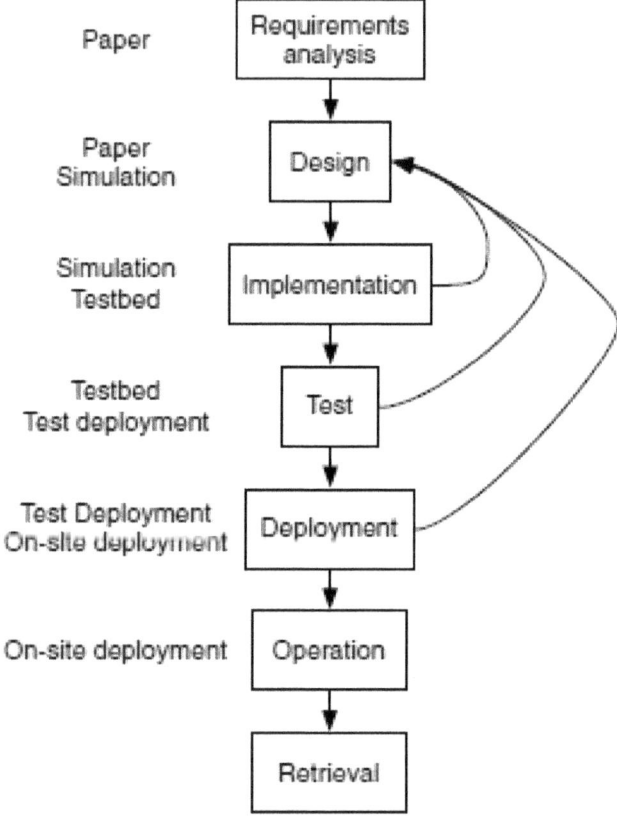

Figure 6.2 Life cycle of WSN

is expected to gather data for humidity, temperature, and solar radiation every 5 minutes. You can assume that these are the requirements from the vineyard owner. Your task is to design and deploy this network as systematically as possible.

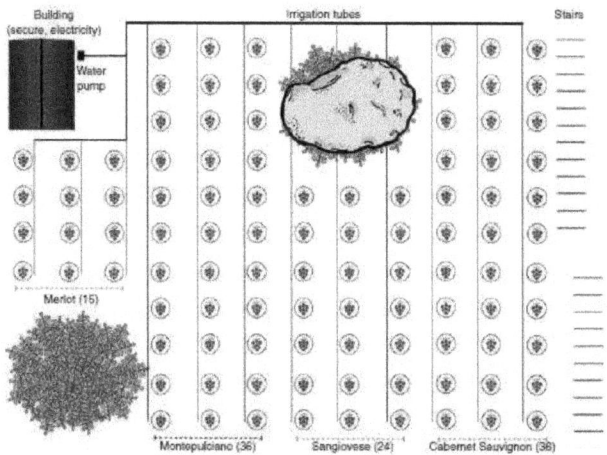

Figure 6.3 Example of Vineyard with WSN deployment

6.4 Requirements Analysis

The main goal of the requirements analysis is to ensure that you understand the application, the environment in which it will be running, and expectations of the user. In this case, the user is the owner of the vineyard who is probably not a technician. However, he or she is an expert in the field of growing and producing wine and expects the new system to simplify and support his or her work. In order to find out how your future application will implement the envisioned application; you need to ask several questions and carefully analyze the environment. The questions you need to ask can be divided into four categories: environment, lifetime and energy, data, and user expectations.

- **Analyzing the Environment:**

This is probably the most challenging and blurry task. You need to analyze the weather and its fluctuations, surrounding flora and fauna, expected interference, and topology of the site itself.

- **Weather:**

The vineyard owner would like to use the sensor network throughout the whole year, even in winter. The owner suspects that the temperature and humidity during winter impacts the quality of the wine in the next season. The vineyard itself is situated in southern Switzerland, close to the city of Mendrisio. Thus, you can expect temperatures from - 20 degrees Celsius in the winter (happens rarely, but happens) up to +50 degrees in the summer time. In terms of humidity, all levels from 0 to 100% can be expected. Furthermore, snow and rain are typical for winter and spring and may last for several days in a row. Full snow coverage is possible for extended periods of time (weeks). This data is best obtained from meteorological stations near the deployment site. *Consequence: Packaging needs to be very robust and the hardware itself needs to be tested in extreme conditions.* **Flora.** During the normal yearly lifecycle, the flora at the deployment site will appear and disappear. In winter, almost nothing will be left while in summer time the vines will be growing. However, you can expect that no other abundant vegetation will be present, such as big bushes or big trees, as those impact the growth of the vines. *Consequence: Nodes should not be buried under the leaves and should have a similar*

Micro-climate around them. **Fauna.** The vineyard is open and not protected from wild animals or dogs. Domestic animals such as cows or sheep are not expected but also not completely improbable. *Consequence: Packaging needs to be robust against wild animals.*

- **Interference:** There are no high-voltage power transmission grids around the vineyard.

In terms of other wireless devices, the only wireless network is the one installed by the owner for home use. However, hiking paths are situated close-by and interference from various Bluetooth or wireless devices can grow during high hiking season. *Consequence: Communications should be carefully tested under high interference and channels with low interference should be selected.*

- **Topology restrictions:** The vineyard is situated on a slope but in general accessible from all sides. However, a big rock is situated in the middle of it, where several vines are missing (Figure 10.3). *Consequence: No sensor nodes can be placed on the rock directly, thus communication around the rock need to be carefully tested.*

- **Security concerns:** These can be both in terms of physical security and cyber security. Tourists typically do not enter the vineyard but it still happens, especially young people at night. The hiking paths are also close so cyber-attacks are not impossible. *Consequence: Robust installation, which does not allow easy removal or displacement of nodes.*

- **Existing systems:** Very often, existing systems can be leveraged for a new sensor network. For example, in the vineyard you see a sophisticated irrigation system, which spans the complete area. In summary, the vineyard application is not extremely challenging, especially because it is easily accessible and weather conditions are not extreme. More extreme environments, such as glaciers or volcanoes, require special measures.

- **Analyzing Lifetime and Energy Requirements:**

Next, you need to understand for how long this installation is supposed to work. Expected lifetime. The vineyard owner needs at least two years of data, the longer the better. Maintenance opportunities. The owner is regularly checking his vines approximately once a week, which will give you the opportunity to exchange batteries at least once every one or two months. Additionally, since the vineyard is publicly accessible, you can easily enter and test the system without the help of the owner (rather a perfect case and rarely seen). Energy sources. Solar batteries are a good option, as sun is abundant on the vineyard, even in winter periods. Given the extended total lifetime, they are probably the better and cheaper option. Next, you need to understand for how long this installation is supposed to work. Expected lifetime. The vineyard owner needs at least two years of data, the longer the better. Maintenance opportunities.

- **Analyzing Required Data:**

Sensory data includes temperature, humidity, atmospheric pressure, wind speed, precipitation, solar radiation, and ultraviolet radiation. These are typical for agricultural monitoring and are all rather low-data rate sensors.

- **Tolerated delay.** The tolerated delay for the data is quite high in winter when no vines are growing and data gathering is only happening for statistical purposes. In this period, even days and weeks are acceptable. However, during vine growing periods, the delay is approximately 15 to 20 minutes so that the sensor network can be directly attached to the irrigation control.

- **Precision and accuracy.** The precision of the data is expected to be normal, e.g.,1 degree Celsius for temperature, 1% for humidity, etc. The data's accuracy should be quite high, meaning that sensor readings should be collaboratively preprocessed to assure that no faulty readings are sent to the user.

- **Time precision.** Time is a critical issue for WSNs because the hardware typically does not have real-time clocks. What is natural to have on a personal computer becomes a real problem to obtain on a sensor node. Thus, you need to understand what the application's requirements are and how much attention you should give to time synchronization. In the vineyard, as in almost any other WSN application, time is important. The time precision, however, is rather coarse and a precision of several seconds to even a minute is acceptable. Locality of data. This requirement refers to the question of whether you need to know exactly where the data comes from or whether this is not important. For the vineyard, it is important to know whether the data comes from node 1 or 51, as the irrigation system will start watering the area around the node. For other applications, e.g., proximity recognition between two people, it is not important to know where location agnosticn the people are, only that they are close to each other. Such applications are called applications *location-agnostic*.

- **Analyzing User Expectations:**

The vineyard owner has great plans for this installation and would like to better understand the topology of the wine fields and which sub-fields deliver best results. The owner would like to lower the use of water and pesticides, and even attempt to transfer to organic production. Furthermore, the owner would like to include later other sensors such as chemical sensors or biological sensors to early identify problems and sicknesses. Currently, testing the system in one of his fields is of interest but optimally the owner would like to extend the installation to all fields and have all data in one place. Thus, scalability and extendibility are important requirements that are best taken into account early.

6.5 The Top-Down Design Process

The design process targets a complete architecture of the sensor network, including where the nodes are, which protocols to use, and many other factors.

However, it does not go into implementation details. Design and implementation are sometimes hard to differentiate.

For example, in the design process the designer could decide to use a particular routing protocol and MAC protocol. However, both protocols are complex and both cannot fit on the sensor node. Is this a design problem or an implementation problem? In fact, it is considered a design problem. The design should make sure that all these details are discussed and evaluated, and that the complete architecture of the system is well analyzed by system experts. Section focuses on the implementation process and deals with connecting individual components and mainly testing them.

The Network:

Given the vineyard and the requirements analysis, you have several options for the general network. You could install a sensor at each vine, having a total of 111 nodes. This is quite a lot for such a small vineyard. At the same time, this deployment will also provide the highest sensing precision. Another possibility is to deploy the network only in half or one- third of the vineyard, e.g., in the half closer to the house. This will provide the owner with high precision and a good point for comparison with the other half. Yet another possibility is to deploy sensors not at all vines, but only at some of them. Here, the experience of the owner is crucial because the sensors need to be placed at the most relevant positions. In this case, these are the places where the land flattens (in the middle of the vineyard, where the stairs are interrupted), and around the trees, as there the land is able to store more water and the temperature is lower. Furthermore, you need to consider that different grape vine varieties (Merlot, Sangiovese, etc.) have different properties and different needs. After a discussion with the owner, it makes most sense to install sensors at approximately every second vine in each direction, as shown in Figure 6.4. The network consists of 36 sensor nodes, each of them gathering temperature, humidity, and light readings. The sensors are positioned to cover the flat areas and different grape vine varieties.

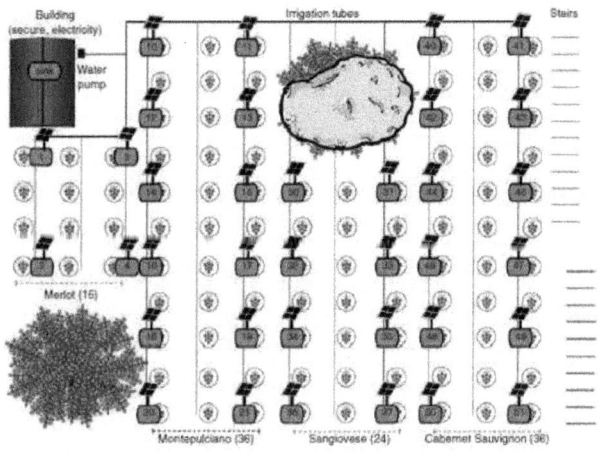

Figure 6.4 Design of Vineyard WSN deployment

Montepulciano and Sangiovese, for example, offer opportunities to see fine-grained differences between the grape vine varieties and between neighboring vines. Instead of considering exchangeable batteries, you should consider solar powered ones as previously discussed. Each sensor will be equipped with a small solar panel, similar to those used for solar garden lamps. The sink is positioned in the house, which is connected to the Internet and power grid. The sink does not need any special protection because it is in a well- protected house with stable temperature and humidity levels. However, the sensors need protection from animals and curious tourists. They will be installed on pillars as high as the vines and firmly connected to them. They are also protected with transparent, but hard packaging and have a QR code to explore. This typically works well, as most of the curious people just want to know what this thing is. A QR code offers a possibility to explore without damaging the sensor node, intentionally or not.

The Node Neighborhood:

In terms of the network topology and the neighborhood properties, since the network is relatively dense and distances short, problems should not be expected. In order to fully secure this, you should perform a couple of experiments in the field and test both short and long links. The problems you are typically looking for are: Isolated nodes. In this case, nodes 41 and 51 might be isolated, as they are the outermost nodes. It needs to be tested whether they are connected to at least two neighbors. Nodes with too few neighbors. This is similar to the previously isolated nodes but less critical. In this vineyard, many nodes can have in theory only one neighbor, e.g., nodes 41, 20, and 51. Chains of nodes. The previous problem often leads to chains of nodes. In this vineyard, the probability of having chains is low because you have installed a grid of nodes. In other cases, chains are very risky, as their entire performance depends on all nodes in the chain and a single failure will break the complete chain.

Nodes with too many neighbors. This problem is quite opposite from the previous ones. However, when a node has too many neighbors, it also needs more energy to maintain information about them and it might serve as a forwarder for too many nodes. In this case, many nodes could have many neighbors, e.g., node 32. This is best handled by the link and routing protocols. Critical nodes. These nodes serve as forwarders for many neighbors and at the same time have few neighbors to reach the sink. In this case, many nodes (1, 3, 12, and 10) are very close to the sink so the risk is low. Otherwise, you would need to provide extra nodes to spread the load.

The Node:

The main node-level problems are dead nodes and seemingly dead nodes. What is important at this stage of the design process is, what will happen if some of the nodes die? If the previous neighborhood-level design was properly done, nothing bad should happen if some of the nodes die. However, you can still differentiate between two types of networks:

- One-hop networks. In these networks, the death of individual nodes does not affect the work of the other nodes. The network can die gradually and even if only one node is left, it can work correctly. Thus, node dying is "allowed" and does not need any special precautions in the implementation phase.

- Multi-hop networks. In these networks, the death of individual nodes affects the work of others and especially their communication. Even if some in-network Cooperation is performed before sending the final result to the sink (e.g., two nodes agree on a temperature value), the network can still be considered a multihop. Here, you cannot really allow for too many nodes to die and you need to take implementation steps to recognize and prevent the death of nodes. In this vineyard, we have a second type of network and thus we need to recognize and prevent node deaths.

6.6 Bottom-Up Implementation Process

The implementation process starts bottom-up implementation process, However, first you need to identify the hardware to use. Essentially, you have two options: off-the shelf or custom-built (home-made) nodes and solutions. For straightforward and rather standard applications we will be often able to find a complete solution on the market, e.g., sensor nodes already connected to the cloud. These look tempting and, in some cases, they might be the perfect solution. However, the following properties need to be evaluated for both off-the-shelf and custom-made solutions:

- **Price:** You can expect the price of a complete solution to be much higher than for home-made solutions.

- **Requirement's fulfillment:** Are all your requirements really fulfilled? You need to check carefully all items in Figure 10.5. For example, some off-the-shelf solutions might not offer solar batteries.

Furthermore, the use of data and their representation might be different from what the user expects.

- **Programmability:** Can you change the software or is it proprietary? This is very important when the application or some of the software components are rather nonstandard. This is often the case with consumer-oriented products, such as smart home systems.

- **Extendibility:** Especially when the user would like to extend the system later, it is crucial to select a system, which can be freely extended and modified. This is clearly given in home-made systems but often a trouble with off-the-shelf solutions.

- **Service agreement:** Off-the-shelf solutions often offer paid service agreements, which guarantee the work of the system to some extent. This is convenient but also inflexible.

If you decide on a complete off-the-shelf solution, you may close this book now. But the opposite decision, making everything from scratch including the hardware is also not in the scope of this book. The typical decision is to select an off-the-shelf hardware, which can be programmed by open-source operating systems and architectures such as Contiki. This should also be your decision for the vineyard application.

Individual Node-Level Modules:

Define first the input and the output of individual modules: For example, for your regular sensing component, the input is a timer firing regularly and the output are these values: temperature, humidity, and solar radiation.

Reuse as much as possible: If existing modules already offer the required functionality, reuse them. If there is already a module for regular temperature sensing, reuse it to add humidity and solar radiation readings.

Ensure you fully understand reused modules: For example, when you decide to reuse the temperature regular reader, make sure it does exactly what you expect it to do. It could be that it only delivers the temperature if it is above some threshold.

Do not make dependencies between the components: This is one of the most important concepts as it allows you to test individual components separately from each other and reuse them for other applications. Do not make any assumptions about which module does what, but implement the module in a self-sustainable way. For example, the link quality protocol can reuse packets arriving for and from other modules, but it should also be able to create its own packets, if no others are present. The testing of the individual node-level modules can be best done in simulation, where many different scenarios and event sequences can be tried out.

Finally, you need to analyze the behavior of the network as a whole. This relates mostly to communication and topology issues and the main concepts are as follows:

- **Reboot handling**: Again, a node reboot might result in unexpected behavior, wrong routing information, time, and location. What happens when a node reboots? How do the other nodes react to this?

- **Communication stands**: Make sure that the implemented communication protocols Work correctly to avoid collisions, congestions, and energy wasting. Testing of the complete network can be conducted again either in simulation or in test beds. However, it is crucial to confirm the implementation in a sample deployment site or in the final deployment site. The choice often depends on costs and the availability of testing possibilities. For example, testing a parking system on a university or industrial campus is feasible, whereas testing the vineyard on a campus is not. Once the individual components are implemented and tested, you should analyze the node as a whole. The main principles are as follows:

- Make sure you have enough memory/processing. Individual components rarely have a memory problem. However, putting all components together might result in a memory and/or processing bottleneck.

- All components are correctly interconnected. Thoroughly analyze the interconnections between the components of the system. Look especially for shared data structures, where reading and writing accesses might create collisions.

- Event order. Sometimes the events in a system might be ordered in a weird way. Consider the following example: perhaps you use two timers, one for reading the sensory data and the other for sending the packet with the data out. However, sometimes the reading timer fires first but sometimes the sending timer does.

If this order is not deterministic for some reason, the packets will sometimes hold duplicate data and sometimes skip readings. The event order is crucial for making the node as a whole work correctly.

•●•

www.ingramcontent.com/pod-product-compliance
Lightning Source LLC
LaVergne TN
LVHW061551070526
838199LV00077B/6993